English Language Learners and Math

Discourse, Participation, and Community in Reform-Oriented, Middle School Mathematics Classes

English Language Learners and Math

Discourse, Participation, and Community in Reform-Oriented, Middle School Mathematics Classes

Holly Hansen-Thomas
Texas Woman's University

INFORMATION AGE PUBLISHING, INC.
Charlotte, NC • www.infoagepub.com

Library of Congress Cataloging-in-Publication Data

Hansen-Thomas, Holly.
English language learners and math discourse, participation, and community in reform-oriented, middle school mathematics classes / Holly Hansen-Thomas.
p. cm.
Includes bibliographical references.
ISBN 978-1-60752-148-8 (pbk.) – ISBN 978-1-60752-149-5 (hardcover)
1. Mathematics–Study and teaching (Middle school)–United States–Case studies. 2. English language–Study and teaching (Middle school)–United States–Spanish speakers–Case studies. 3. Hispanic American children–Education (Middle school)–Case studies. I. Title.
QA13.H36 2009
510.71'2–dc22

2009021553

Printed in the United States of America

For my family,
little Margitta Luise, Sadie Lina,
and their Papa Billy

CONTENTS

ACKNOWLEDGMENTS

This work could not have been completed without the support and guidance of some terrific people and organizations. First and foremost, I am extremely grateful for the participation of the teachers and students at Hot Springs and Ritter Middle Schools. For this, I must also express my appreciation to the administration of the schools and to the parents of the kids I had the privilege of working with during their sixth grade year. ¡Muchissimas gracias a todos!

Next, I must thank all of my advisers who read so closely the original manuscript from this work (my dissertation) and helped me to think about ways in which I ultimately took to better it. In particular, I thank Juliet Langman and Bob Bayley for their genuine attention and care. I am also highly appreciative to my writing group colleagues at Binghamton, and to anonymous reviewers for all of the comments, questions, additions, subtractions, and suggestions for improvement. It goes without saying, though, that I myself am responsible for all errors and omissions.

A dissertation writing grant from American Educational Research Association-Institute of Education Sciences (AERA-IES), and another from Kappa Delta Pi (KDP) supported my writing of the original version of this manuscript. I thank the 2006 National Association for Bilingual Education (NABE) Outstanding Dissertation Award Committee for their kind comments and recognition of the original version of this work. Students at both Binghamton and TWU helped by checking references and citations, and I appreciate all of the detail work.

Finally, I could not accomplish anything without my dear family: baby Maggie, big sister Sadie, and my wonderful partner, Billy Thomas. I am also indebted in gratitude to my parents, Al and Sheila Hansen, who so enthusiastically supported this project, as well as many others! Thanks to all!

English Language Learners and Math, page xi

INTRODUCTION

This book is written for education professionals, as well as parents, friends, and others interested in the educational experience that middle school English Language Learners (ELLs) in mainstream content-area classes face. Taking a community of practice (CoP) perspective that highlights the learner as part of a community, rather than a lone individual responsible for her/his learning, this study investigates how six Latina/o ELLs in three middle school math classes negotiated their learning of mathematics and mathematical discourse. The classes in which the six Latina/o students were enrolled used a less traditional approach to math learning; the math in these classes was—to varying degrees—taught using a hands-on, discovery approach to learning where group learning was valued, and discussions in and about math were critical. The three sixth-grade teachers I report on here fell along a continuum with respect to their use of non-traditional, or reform-oriented math. Of the three teachers in Springvale District, a large urban school district in a metropolitan community in the Southwest, Ms. Koch used it to a great extent, Ms. Loesely to a moderate extent, and Mr. Martinez, to a small extent. Thus, some of the ELLs had more experience learning math in traditional ways, while others learned using a reform-oriented approach to math learning.

Part of my interest in carrying out this study stemmed from my prior experience as a middle school English as a Second Language (ESL) teacher. When I was teaching ESL, I found myself playing defense (as I often did) for one of my more troublesome, but exceptionally charming, eighth grade boys from the Middle East. This boy, called Lalo, by the Spanish speakers in ESL class (since his real name was similar to this name that was familiar

English Language Learners and Math, pages xiii–xiv

to them), was sharp, popular (at least in ESL class), and very energetic. He also had a very fragile side, having come to the U.S. with only his father. They had left his mother and siblings back home until they could raise sufficient funds to bring them to the U.S. Sometimes for Lalo, school was not his first priority. One day, the eighth grade Algebra teacher (and sports coach) reproached Lalo for not performing well in class—because, according to the teacher, math is a 'universal language'. Talking to Lalo, I discovered that his teacher expected him to follow a very specific format of solving problems in math. Lalo explained that he had never learned the (American) way of solving algebra problems that his teacher expected of him. I knew that I, as the ESL teacher, had to alert the teacher to challenges that Lalo, and other ELLs in his classroom were experiencing. This issue, coupled with an increased emphasis on mathematics in today's educational landscape, provoked me to explore the problems that Lalo (and others like him) encountered in mainstream mathematics classes.

My perspective is also shaped by the fact that I am from the same large, diverse, bilingual and bicultural state that the focal students in my study lived in. I speak the focal students' L1 (first language), Spanish, but I am not a native speaker, despite the fact that as a child I spent considerable time on the U.S. border of Mexico with native Spanish-speaking relatives. While I certainly do not profess to share the same background as the students I worked with in my research, I have some comprehension of their experience.

Without question, my history and experience have shaped my perceptions of how language learners learn their L2 (second language) in school. Moreover, my background contributes to my conceptualization of 'appropriate' and 'inappropriate', and 'effective' and 'ineffective' teaching and teachers. However, I attempt to maintain objectivity as I present the stories of how three teachers and six language learners worked with each other to gain a degree of competence in the mathematical discourse they used in class. As a qualitative educational researcher, I acknowledge that the decisions I make with respect to analysis of teachers, students, and all collected data are colored by my training, education, philosophy and background. I thus acknowledge that I am 'constructing a reality' (Merriam, 1998, p. 22) based on my observations, chosen theoretical frameworks, and analyses.

GUIDE TO TRANSCRIPTION CONVENTIONS

Based on CHAT (Codes for Human Analysis of Transcripts), transcription and coding system of CHILDES (Child Language Data Analysis System) (MacWhinney, 2000)

xx/xxx -unintelligible speech (number of xs may correlate to words)

[?] -best guess at a word

wanna [: want to] -assimilation
cuz [: because]

+/. -interruption

+... -trailing off

<talk> [>] -overlap follows

<talk> [<] -overlap precedes

or ## -Pause (depending on length or number of seconds)

— -classroom talk omitted/more than 10 seconds elapsed between segments

CHAPTER 1

BACKGROUND

The face of U.S. education in the twenty-first century has changed tremendously in recent decades. The student population has become increasingly diverse, with respect to place of birth and languages spoken in the home. Although U.S. educational institutions have passed myriad immigrants through their halls over the years, today's largest immigrant groups are quite different than those in the past. The majority of immigrants in schools are no longer of European descent as before, but hail from Latin American and Asian nations (Echevarria & Graves, 2007; Molesky, 1988). Of the total foreign-born population in 2003 of more than 33.5 million in the U.S., slightly more than half (53.3%) came from Latin America (American Community Survey, 2005; U.S. Bureau of the Census, 2000, 2003). Further, of the 4.3% of foreign-born students (children under 18) in 2005 in the U.S., 53% were Latina/o, and 20% were Asian (Institute of Education Sciences, 2007). Moreover, many new immigrants, as well as first, second, and sometimes even third generation school children, come to school as English language learners (ELLs), having spoken a language other than English in the home until entry in school. Almost 18%, or 1 in 6 people five years of age and older come from a household that speaks a language other than English (Bayley, 2004; U.S. Bureau of the Census, 2000). ELLs represent more than 10.5% of the total K–12 population (National Clearinghouse of English Language Acquisition (NCELA), 2006), with a slight majority in elementary schools (52.5%) (Capps et al., 2005). There are also quite high numbers of ELLs in U.S. mainstream secondary content-area classes (Batalova, Fix, & Murray, 2005).

English Language Learners and Math, pages 1–6

Changes in student demographics have emerged at the same time as changes in student performance. Student failure in important content area classes such as math, science, and reading has paved the way for educational reform. In math, current curricular reforms are characterized by an emphasis on problem solving, understanding concepts (instead of just algorithms) critical thinking through discovery learning, and communication and discussion in group situations (National Council of Teachers of Mathematics (NCTM), 2000). In today's world, students must be able to not only to compute and calculate—in the case of math—but also to be able to talk in and about math in order to succeed in school.

The intent of this research is to examine in detail the mathematical learning situation of a particularly important group of ELLs in U.S. schools today—Latina/os. The present study aims to understand how Latina/o ELLs—(including relatively advanced English-speaking bilinguals (who were born in either the U.S., or who arrived at a young age, i.e., early elementary school), and newer, recent arrival immigrants (having arrived later in elementary school)) develop and learn to use math discourse in school in the wake of recent educational math reforms.

The primary question guiding this project asks: How do language learners develop math discourse and gain math content knowledge in a reform-oriented classroom? Since a focal point of this investigation centered around the notion of community of practice, or CoP, I also wanted to know what the role of participation was in the learning of the students I followed—an important component in the CoP. Thus, I was also seeking the answers to the following questions—in hopes that they could help me gain a better understanding of how Latina/o ELLs learn math in reform classes.

- What is the effect of participation in the classroom community on the successful use, development of, and socialization to math discourse and content knowledge?
- How do individual variables such as language proficiency, gender, knowledge of math and the 'ways of school', and academic identity affect participation in math? and
- What features are present in reform-oriented math curricula, or those classes which utilize this curricula, that facilitate learning of mathematical discourse for ELLs?

METHODOLOGY AND DATA SOURCES

The primary methodology I use in this study is a qualitative, case study format with an ethnographic perspective. I analyze the communities of practice as they emerged within natural student-produced classroom discourse in the three mathematics classrooms I followed. The format of discourse analysis used here is that of interactional sociolinguistic discourse analysis, credited to Goffman (1983) and Gumperz (1982). Therefore, the data sources that are considered in this project include (audio and video-recorded) classroom discourse; classroom observations; artifacts including student-produced work and tests; student records such as report cards, standardized test results; and information regarding prior math and language placements and achievements from six students, their teachers, and their classmates from three schools in the same school district. Fieldwork was carried out throughout most of the academic year, 2003–2004, and the majority of the concrete data used in my analyses came from the second semester of the school year (January–May). In order to better understand the background, experiences, and motivations of the ELLs, I conducted two sets of one-on-one interviews with each focal student (one early in the semester, and one at the end) as well as interviews with key (non-focal participant) group members to better understand focal students' classroom identity. I also spoke formally and informally to all three participating teachers to gain insight as to their philosophies about using reform math and working with second language learners. These conversations were either audio-recorded or recorded as field notes.

I chose the classrooms based on their use of reform-oriented mathematics. I asked local mathematics professionals (teachers, administrators, professors) which schools were using reform math, and then proceeded to obtain permission to work with several experienced and highly regarded teachers. All of the teachers used a particular type of reform mathematics curriculum, the Connected Math Project (CMP), to some degree in their sixth grade math classes. They were mandated by the district to use this curriculum, but were not required to use it exclusively. As a result, the three teachers differed in their use of CMP.

CMP was developed through a grant by the National Science Foundation and is characterized by a focus on connections, problem solving and sense-making, discussing and verbalizing, and generalization of patterns and relationships (Lappan et al., 1996). Thus the practice of CMP involves math learning through group work, use of manipulatives, and hands-on learning in a thoughtful, global, and discursive way. Cooperation and interaction are then integral components of CMP.

SOCIOCULTURAL PERSPECTIVE

A sociocultural perspective provides the broad lens through which I view and operationalize the research questions addressed in this work. A sociocultural theory of the mind maintains that human mental activity is mediated through "symbolic artifacts to establish an indirect, or mediated, relationship between ourselves and the world" (Lantolf, 2000, p. 1). One very important symbolic artifact that humans use to mediate human mental activity is language. To L. S. Vygotsky, the progenitor of sociocultural theory, mediation through both language and social interaction is a necessary requirement for higher mental functioning (Vygotsky, 1978). Explicit also in Vygotsky's conceptualization of a sociocultural approach to the mind is the idea that mental processes are developmental (genetic), and should be studied and analyzed as such. Of the genetic domains he identified for which to study mental abilities (phylogenetic, sociocultural, ontogenetic, and microgenetic), the ontogenetic domain of moment to moment interactions has been the most studied, by not only Vygotsky himself, but also by other sociocultural researchers (Lantolf, 2000; Wertsch, 1991). Wertsch and Lantolf's research centers in particular on language development. Following this tradition, the research described in this book will be carried out within the ontogenetic domain, in which the focus of the study "is on how children appropriate and integrate mediational means, primarily language, into their thinking as they mature" (Lantolf, 2000, p. 3).

Vital to Vygotsky's (1978) conceptualization of learning is the ZPD, or zone of proximal development. The ZPD is defined as the distance between one's actual developmental level (without assistance, or scaffolding, by an adult or more capable peer) and the learner's potential level of learning when they are scaffolded. The Vygotskian view holds that learners must be in this zone in order to develop knowledge. Cummins (1994) provides a succinct representation of the ZPD as "the interpersonal space where minds meet and new understandings can arise through collaborative interaction and inquiry" (p. 45). According to sociocultural theory, it is this space where learning occurs. I take the view that ELLs participating in mathematical communities of practice have the potential to develop linguistic and mathematical knowledge from one another—thereby interacting in the ZPD in the math classroom.

DISCOURSE

In this study, I examine use and development of mathematical discourse by ELLs. I conceptualize the use and development of a discourse—also referred to in the literature as a register, language, or literacy—as a tool for learning as well as an outcome. The relationship of discourse to learning is thus as both a product and a process. In this study, I have chosen to emphasize the role of discourse as a process, but also examine the product to understand what students learn in math.

Gee (2005) contrasts two types of discourse: Discourse (big, or capital D) and discourse (with a lowercase d). The latter type, or 'discourse' is defined by Gee as language in use such as stories or conversations (p. 26). The former, or big D Discourse is, according to Gee, "language plus the other stuff" (p. 26). He goes on to describe that big D discourses are linked to identity and activity and that they are "embedded in . . . social institutions . . . such as . . . books . . . and classrooms" (p. 27). Thus, for the purposes of this book, the term 'discourse' will be used similarly to the way that Gee utilizes big D Discourse.

COMMUNITY OF PRACTICE

The community of practice (CoP) framework is a learning theory that is rooted in the assumption that "engagement in social practice is the fundamental process by which we learn and so become who we are" (Wenger, 1998, p. i). The trinity of learning, meaning, and identity are critical to the community of practice learning theory. No less important in this framework is the notion of apprenticeship. In the community of practice, a learner's identity evolves and transforms from that of peripheral participant in a specific practice, such as that which develops in math class, to a legitimate peripheral participant, to a potential full participant. Expert guidance, or scaffolding, to use Vygotskgian terminology, is a requisite feature in the learning process as novices are 'cognitively apprenticed' by more experienced peers (Rogoff, 1990).

In my study I look at how both teachers and more capable peers work to apprentice the focal students I follow. What mustn't be overlooked in a community of practice is that both more and less experienced participants work together to socialize each other to the new content or skills to be learned. Learning within a community of practice is thus a two-way process.

With this said, however, I acknowledge that the CoP has been criticized for several reasons. First, it has been noted that CoP is not an effective lens for which to examine learning of language minority students as it doesn't take into account the complexities that ELLs experience in school (Kanno, 1999; Rampton et al., 2002). In particular, the research maintains that in some cases, ELLs are prevented from interacting with native speakers, and are then unable to attain the desired outcome of the CoP model: becoming full participants in the classroom community (Kanno, 1999). CoP has also been criticized for ignoring coersive power relations between ELLs and the institutions in which they are positioned (Varghese et al., 2005). I acknowledge this limitation and concede that conflicts between home and school cultures may arise with ELLs learning math within a reform-oriented framework. Participation and engagement (within a common 'practice' like a math class)—two crucial components of CoP—can in fact be obscured by issues of power and ideology when school promotes one way of learning that conflicts with that which is valued at home. But because I ultimately found that a variety of forms of participation (i.e., English or Spanish use; oracy, gesture, or literacy) were valued, and indeed given status in Mr. Martinez', Ms. Loesely's, and Ms. Koch's math classrooms (but to differing degrees, depending on the teacher, the day, or the task), students usually had sufficient opportunities to demonstrate their integration into and participation in their classroom learning communities. As a result, I found the CoP framework to be appropriate for this study.

CHAPTER 2

INTRODUCING THE THREE COMMUNITIES

Sfard's (1998) participation metaphor (which is contrasted with the more static acquisition metaphor) describes how learning occurs through active participation in communities. This notion plays a critical role in conceptualizing how learning occurs within the present study. To gain an understanding of the macrolevel processes that inform the classroom dynamic and the discourse development ongoing in the class, I draw upon the CoP framework. As mentioned, the CoP framework posits that learning occurs as participants transition from legitimate peripheral participants to full member-participants in a particular community of practice. Explicit in the CoP is the notion that social interaction facilitates the construction of knowledge (Wenger, 1998). Therefore, tenets of social constructivist and sociocognitive theories undergird the CoP framework.

This chapter will introduce the participants in each of the three classroom communities I followed. I give a general description of the three classroom teachers with respect to demographics, personality, teaching style, and interactions with students. I then provide a brief overview of each of the students I included in my study. I describe how these students came to be focal participants in my study as well as other characteristics that define them.[1] I address issues of language and their experience with Mexico, their attitudes toward school and standardized test scores,[2] and provide an introduction to how each of these students interacts in math class. Later, I

English Language Learners and Math, pages 7–19

will paint a picture of how the identities of the six students, along with their respective classroom teachers and assistants worked in concert to create community and facilitate learning for some.

TEACHERS

The three teachers included in my research study represent different approaches to the teaching of mathematics. While they were all veteran teachers, and all came recommended by their administrators and/or local mathematics specialists, their backgrounds, teaching methods, and their outlook on ELLs and reform mathematics teaching were different. Hence, comparisons of their approaches will allow a richer understanding of the complexity of learning math in middle school.

Mr. Martinez

Mr. Martinez had considerable teaching experience. A longtime certified mathematics teacher, he had been teaching for seventeen years. He was a contractor for over a decade and made it clear to me that he returned to teaching due to lack of work in the construction business. He also explained that he spent his free time—outside of teaching and running a daily after school program for students needing extra help in school—as a handyman. He was proud of his carpentry skills, and he made reference to his time in the construction business often in class. Mr. Martinez also indicated that he kept so busy because his wife wanted him out of the house (that he built for her). He lived very close to the school in which he worked, Hot Springs Middle School, and all of his grown children had attended Springvale Independent School District (SISD) schools.

Mr. Martinez, a well-pressed, slightly graying, later middle aged Mexican-American fluent in both Spanish and English, used Spanish often to help meet the needs of the ELLs and bilinguals in his classes. He had found several middle school level bilingual reference texts which listed mathematical terms and phrases in Spanish, and used these books himself to express the appropriate terms in Spanish to the students. While the majority of his discourse throughout the time I spent in his classes (from November to May of a full academic school year) was in English, he regularly used some Spanish for clarification to help out two newcomer sisters from Mexico who were both in second period math. On our very first meeting, Mr. Martinez said that he felt Mexicans were generally advanced in math. He cited the fact that crewmembers from his time as a contractor knew the Pythagorean theorem and would put it to use on the worksites.

Mr. Martinez' teaching style was traditional—in terms of classroom set-up. He conducted his classes in a teacher-centered format, often using the overhead projector, or occasionally the chalkboard to facilitate lessons. Students were seated in rows facing the front of the room, and most classes were carried out as a whole group. He would call on individual students for answers, or more often, students yelled out responses to the teacher's queries. After conducting a whole class lesson, Mr. Martinez would often give students some type of written task, such as a worksheet or book problems to be completed in the remainder of the class or for homework. Many students in class would either copy off another student or put the assignment away and would talk about social matters until the bell rang. Most days, Mr. Martinez would conclude the lesson at approximately five minutes until the bell, inviting noise and off-task talk. Strict discipline and classroom organization were not high priorities for Mr. Martinez. He was somewhat lax with regard to classroom management.

During class, he usually allowed students to leave class to go to the bathroom, so there was often a steady stream of both girls and boys handing off the Martinez hall pass, as though it were a baton in a race. Class rules dictated that no makeup—for boys or for girls[3]—was allowed to be applied in class, however, there were often small, surreptitious groups of girls at the back spraying perfume or putting on lip gloss. He disciplined only the worst offenders of this rule.

Mr. Martinez' interactions with the students were characterized by jokes and teasing. He playfully picked on students that he knew could "take it" and relied on the clever and enthusiastic students to participate in class discussions. He liked his students and they liked him.

Regarding CMP, Mr. Martinez frankly revealed to me that he was not fond of CMP and did not use it as often as he was asked to do by the district math coordinator. He explained that he did not feel comfortable using the manipulatives and that he felt the curriculum did not provide enough practice in basic skills. He did use it minimally, but he also relied on tested and true consumable worksheets and the state adopted Glencoe mathematics textbook (Boyd et al., 2001).

Ms. Loesely

Ms. Loesely was a self-professed non-traditional teacher. She often cited the fact that she gained her certification to teach through an alternative program,[4] although she had been teaching in private, public and elementary and secondary schools for more than 23 years. Sporting a brown bob with bangs, Ms. Loesely dressed in ethnic as well as smart, fashionable, and somewhat offbeat clothing—in spite of her sixty plus years. A mother of several grown children who had been educated at a private magnet school throughout their school years, and now hold postgraduate degrees, Ms. Loesely valued education, and was very enthusiastic about participating in my study. After beginning school years earlier in another Southwestern state, she earned her B.A. degree in business from a local university, once her children had completed high school. Although she truly loved the teaching profession, she expressed to me quite often that she had 'retirement fever' and was looking forward to the day when she could move out of state to be near her children and grandchildren. She had been teaching in the district for nearly a decade, but had transferred to Hot Springs Middle School after teaching at another school in the district. In her earlier years in the district, Ms. Loesely had conducted numerous math workshops and in-services both locally and at international conferences. Some of these workshops had been carried out with Ms. Koch, from Ritter Middle School. Lately, though, responsibilities had eaten up most of her free time, as she was caring for an elderly parent and teaching full time.

In spite of her many years in the Southwest, Ms. Loesely, an Anglo, did not speak much more Spanish than *andale, arriba, por favor,* and *gracias,* although she felt that her receptive skills were much stronger. She did, however, use the limited Spanish she knew with Spanish dominant students. To support the ELLs in her class, Ms. Loesely encouraged peer tutoring and group work. She was intentional in her grouping, and tried to match students weak in math and English with bilinguals with strong math skills.

Her teaching style was eclectic. Sometimes Ms. Loesely would present lessons in a lecture style format, having the students arranged in traditional rows, but more often than not, she would group the students in pairs or small groups and encourage them to work together. She generally advocated a 'non-traditional' classroom, in that she promoted games, the use of hands-on activities and manipulatives involving scissors and construction paper, dice, blocks, food, and other sensory realia. The students in Ms. Loesely's class were usually busy from bell to bell. Typical warm-up activities[5] involved copying and solving a problem from the overhead projector and then discussing it as a class. The default classroom arrangement was rows facing the teacher, but she modified this according to the constraints of the daily lesson.

Ms. Loesely's class rules were not uncommon for math class. Students were not allowed to use pen for writing assignments, only pencil. If a student did not have a pencil or could not borrow one from a fellow student, he or she had the option of paying five cents, or by taking lunch detention for a pencil from the teacher. Students were expected to write a proper heading on their papers and turn them in on time. She demanded that unacceptable or incomplete assignments or exams be redone. If students forgot their homework or do-over assignments, they would have to come in at lunch to make them up. Students weren't usually allowed to leave class to go to the restroom, unless they had a good reason for doing so. In fact, she was so adamant about this rule that on one day, when nearly half of the class requested to go to the bathroom, she decided to personally escort more than twelve students to the bathroom at once. When students would slouch in their chairs, she would admonish them with the reminder "This isn't Club Med, sit up!" However, Ms. Loesely altered her rules and regulations according to the activities they were doing in class. Late in the semester, when the class had begun working on visual spatialization, Ms. Loesely indicated that "the Club Med slouch is okay today" because the students needed to be able to see different configurations of shapes at different angles.

Ms. Loesely was quite comfortable with her students, and although they saw her as tough, she was fun and made math enjoyable. She would often have previous students come by her classroom to visit with her between classroom breaks. Because she taught the only honors sixth grade math class[6] she was known throughout sixth grade as the teacher for smart students.

While she generally liked CMP, and used it much of the time, she felt that as a curriculum, it had its advantages and disadvantages. The advantages, according to Ms. Loesely included the use of the realia, the open-ended activities, and the group work. Disadvantages included its lack of practice activities. "You can waddle and wobble through it" she said, but exclaimed that something is missing from the curriculum. "It's playing, not practicing" she said. On another occasion, she maintained that it was very hard to give grades in CMP, with all the group work. Citing specifics, Ms. Loesely remarked that within the fraction unit, there were no practice activities on dividing fractions by numbers other than ten or a hundred. She also believed that when students had used the holistic, reform math elementary curriculum prior to coming to middle school, that they did not have the basic skills they needed as sixth graders. In addition, she felt that the urban middle school kids she worked with lacked the background needed for a complete understanding of the scenarios presented in the CMP texts. She admitted that her lack of comfort with some aspects of the curriculum derived from the fact that the school year prior to the one in which I collected data was the first time she had ever used the sixth grade CMP materials. She was a seventh grade teacher in her previous years at Hot Springs and had used CMP for seventh graders.

Ms. Koch

According to the principal of Ritter, Ms. Koch was one of best teachers at the school. A tall, honey-haired woman with a beaming smile, she had been honored by a local educational organization for excellence in teaching, and was active in mathematics organizations such as the National Council of Teachers of Mathematics (NCTM). She often presented at and attended these and other math conferences, as well as presented at local in-services and workshops with Ms. Loesely and other colleagues. In her forties and a good deal younger than both Mr. Martinez and Ms. Loesely, Ms. Koch had taught for seventeen years, beginning at a parochial school, and then going on to public elementary and then middle school. Originally from the deep South, she moved to the area as an adult after a tour of duty abroad in the military. She was very proud of her bright daughter who was on scholarship away at college. Ms. Koch often spoke about the many international places to which her daughter had traveled. Ms. Koch proudly showed off photographs taken of her daughter in the same places and poses in which Ms. Koch herself had had photographs taken. She was also very proud that she had taught an up and coming rookie professional basketball player who was something of a local hero. She often included him in teacher-produced activities such as warm-ups and tests and the students were very aware of and proud of Ms. Koch's relationship with the young ball player.

Although her time in the service exposed her to many European and Asian cultures and languages, Ms. Koch did not speak Spanish, the L1 of the majority of her ELL and bilingual students. She did, however, have a Spanish/English bilingual assistant in her class, Mrs. Hammer, as well as a bilingual student teacher, Mr. Acuña. Both of these assistants were often in the classroom with Ms. Koch and served as additional support for the teacher and the students. Ms. Koch had also picked up several advanced mathematical Spanish/English dictionaries at her most recent NCTM conference. She explained that she would allow ELLs to use these kinds of references in class, if they needed, although I did not witness the students' use of these books.

In terms of classroom management, Ms. Koch ran a tight ship. Students were expected to be quiet and on task, unless instructed otherwise. Students sat in groups with four to five students at each table. Classes began with a daily warm-up activity to copy from the overhead projector and solve on paper. Ms. Koch then reviewed the task in a whole class discussion format, students passed up their daily warm-ups to one person at their table, and she would then collect the papers from the designated paper collector. Then she would introduce the daily lesson in a whole class format and then usually put students to work on a task in their groups. At this point students were free to discuss mathematical matters.

Ms. Koch's classroom rules were straightforward. There was no talking when the teacher or another student was talking (except during group work). Students did not leave their seats unless they received permission to sharpen a pencil, or throw away trash, nor did they leave the classroom unless they were requested to do so. Ms. Koch taught the students various learning strategies such as guess and check, find a pattern, make a list, etc., and expected the students to follow these strategies as they solved problems in class. She had very high expectations of her students and they were aware of this. I am sure that she would have disciplined the students if they talked out of turn, or got out of their seats, but in the time I spent in their classroom, they never disobeyed the rules.

Her serious demeanor promoted the reputation of Ms. Koch as a 'mean' teacher. However, she was highly respected by the students throughout the school. In fact, one of the main office secretaries once informed me that her daughter had Ms. Koch as a sixth grader, and she reported that it was the hardest, but probably the best math class she had ever had. Ms. Koch was an advocate of rewarding students for good behavior. Some days, when students would correctly answer a question, she would surprise them with lollipops. Another time, she promised students who made a ninety or above on a state standardized practice test a soda water to drink in class. These surprise rewards served to strengthen bonds between Ms. Koch and her students.

With respect to the curriculum, Ms. Koch was a strong proponent of CMP. Not only had she been trained extensively in using the curriculum, but she had taught the lessons forty to fifty times in her past seven years at Ritter. Due to her familiarity with the curriculum, her organizational skills, and her clout with teachers and parents, Ms. Koch had arranged for all sixth-grade students on the team for which she served as math teacher, to complete several CMP books in advisory—the study period which met the first hour of every day. As a result, none of her classes were rushed to complete the series of CMP books. She had enough time to cover the material they needed to complete, and also had the time and freedom to integrate additional curricula to fill in any gaps left by CMP.

I asked Ms. Koch if she had any problems or complaints about CMP, but she was unable to come up with specifics. When I mentioned Ms. Loesely's and Mr. Martinez' complaints that CMP did not provide enough practice, was hard to grade, required background knowledge the students didn't have, and used too many manipulatives, Ms. Koch countered that she used additional activities to fill in the gaps that CMP left, and took grades on warm-ups and group activities. She conceded that manipulatives were more work, but felt that the benefit they brought students were worth the effort. With regard to the issue that CMP requires background knowledge that students in SISD wouldn't have, she felt that minimal scaffolding by the teacher or more capable peers would invalidate the issue. In the next section of this chapter I provide brief descriptions of the focal student participants included in my study.

FOCAL STUDENTS

Thomas and Benjamin

I chose Thomas, age eleven, and Benjamin, age twelve, to focus on in Mr. Martinez' class. A very interesting pair in a somewhat chaotic classroom, these two boys sat very near each other in Mr. Martinez' class. Throughout the spring semester, Thomas and Benjamin were seated near each other at the front of the classroom near the door. Thomas headed the first row of seats, and Benjamin sat in the next row, directly behind Thomas. Although they did not work in groups on structured activities, Mr. Martinez often allowed the students to leave their seats to help their peers. Thomas usually finished his written work before anyone else did and if he wasn't summoned by his friends, he was enlisted by Mr. Martinez to help others in need. It was normally Benjamin or another nearby friend to whom he gave aid, but because Thomas was known as the 'smart kid' in the class, students who on some occasions made fun of him, called on him for help when they needed it.

As a result of his fair complexion, his almost exclusive use of English in class, and his name (Thomas Trout), Mr. Martinez was unaware that Thomas spoke mostly Spanish at home and had been in bilingual classes throughout his elementary school career. Thomas was small for his age, had a light complexion and often wore his light chestnut hair in a military-like buzz cut. Thick soled tennis shoes formed an integral part of Thomas' school attire, highlighting his too-short, ankle length pants.

Standing about a head taller than his friend, Benjamin wore his black hair in a similar buzz-cut as Thomas. Benjamin had a broad smile which accentuated his round face and slightly chubby physique.

They were both bilinguals—Benjamin was a recent immigrant from Mexico, having arrived in the U.S. a year and a half to two years prior to data collection, and Thomas was born in the U.S. At home, Benjamin spoke only Spanish. Thomas always spoke Spanish with his mother, and both languages with his father. In school, though, both boys almost always used English.

When I asked Thomas what language he used to talk to and help Benjamin in math class, Thomas responded: "English, cuz he wants me to talk to him with nothing but English." Since Thomas' dad ran an automotive mechanic shop on two sides of the Mexican border, Thomas' family traveled to Mexico several times a month, summers, and most holidays. Although Benjamin still had considerable family in Mexico, he said his family didn't return often because of "the pass." The Homeland Security Act, which had recently been enacted, made it much more difficult than it had been in the past for new immigrants to cross the U.S.–Mexico border.

Both boys admitted that they liked school, but Thomas was the only one whose favorite subject was math. About the class and the standardized exams, he explained: “It was easy!” In fourth and fifth grade, Thomas performed well on the state standardized tests. Both years he met minimum expectations, but was awarded ‘commended performance’ in math in the fifth grade.

Contrasting the educational system in Mexico with that in the U.S., Benjamin, the recent immigrant, felt that school was different in terms of size (in Mexico the schools are much smaller), equipment (Mexican schools have fewer computers), classes, because there are no electives in Mexico, and content difficulty. His experience as a fifth grader in an American elementary and as a sixth grader in middle school led him to believe that in Mexico “they didn’t show us that much, like a lot, and on here, they do, and on elementary they did too.” When I confronted Benjamin with the oft-cited notion that mathematics is more advanced in Mexico, he disagreed, and gave the rationalization that “they (Mexican teachers) didn’t show us that much math on the like fractions, and, like that.”[7] As a fifth grader, Benjamin took the state standardized test in Spanish, and met minimum expectations in math, but did not meet the minimum in either reading or science.

Thomas’ penchant for answering the teacher’s questions and making 100’s on his assignments gave him the ‘know it all’ moniker. On more than one occasion, I heard Thomas’ classmates suggest he be sent to Ms. Loesely’s class—implying he wasn’t fit for the class he was in. Outwardly, he did not seem to be fazed by the teasing. He seemed proud that his classmates consulted with him when they needed advice or help with their math assignments. In contrast to Thomas, Benjamin was generally quiet in class. He did, however, seem comfortable participating in noisy class discussions when he was not likely to be called upon individually.

Nestor and Jennifer

I focused primarily on Nestor, age eleven, and Jennifer, age thirteen, in Ms. Loesely’s class. Both slightly chubby, Nestor and Jennifer were very sweet and personable kids. Jennifer normally wore her long, dark-brown hair in a ponytail, and at sixth grade, was begrudgingly getting used to wearing glasses. Nestor was often seen sporting a red wind/rain jacket, no matter the weather.

Since Ms. Loesely often changed her seating plan, students were often shuffled to different parts of the classroom. Because he was quiet and needed extra attention and help, Nestor was usually positioned in one of the very front rows of the classroom, near the door. Jennifer was often seated in a row opposite the door, but near the front of the room. She usually had someone near her that she could go to for help with linguistic or mathematical issues, like her close friends, bilingual classmates Patricia, Kristina, or Yolanda.

Nestor had been in bilingual classes throughout elementary school, but since his family moved so often, his academic records were scarce. Now, in sixth grade, he moved to ESL classes. Jennifer had just been exited from ESL at the end of the fall semester of the current school year, and had spent all of her elementary years in bilingual education. Because she was a 'recent-exit' (from language support services), she was still being monitored by the school.

Both Nestor and Jennifer were American-born citizens, with strong family ties to Mexico. Nestor indicated that he spoke Spanish exclusively with his parents, who came from northern Mexico. He also explained that he had attended summer school in Mexico during school vacations. Jennifer spoke English with her mother and sister, but spoke Spanish with her grandparents and aunt, with whom she was close, despite the fact they lived two hours away in a busy border town.

Jennifer, who considered herself an average student, met minimum expectations in standardized tests in fourth and fifth grade. Despite attending bilingual classes for the same amount of time as some of his classmates, shy Nestor was exempt[8] from taking standardized tests even in Spanish, as a result of his status as a Limited English Proficient, or LEP student. He nevertheless had strong opinions about math. Nestor felt that the math he was doing in sixth grade was different from other kinds of math "cuz you, you gotta 'splain your answers and . . . restate, plan, work, and check." Math in Mexico was different, because there, Nestor explained "we don't have to 'splain nothing . . . you do your stuff by yourself, and they got you like, here's a little kid and here's another, and that's why they don't tell you to do nothing."

Jennifer was often an enthusiastic participant in daily math lessons. Although she did not always produce the correct answers, Jennifer usually stayed on task and carried out the teacher's directives. She enjoyed working with her (girl) friends in math class, but sometimes Jennifer needed to get help with math homework at home. Because, according to Jennifer, her mom equated math with "Chinese writing," Jennifer had to get help from her older sisters. In contrast to Jennifer, Nestor was usually very quiet in math. When working in partners, though, Nestor did interact and participate. Although each of these two students' participation in class was rather disparate, they both succeeded in making themselves known to the teacher and to the rest of the class.

Anna and Pedro

Ms. Koch chose Anna and Pedro as research participants for me. Ms. Koch felt strongly that her second period class would be the best for me to work in, but I was dismayed when I learned that Anna was the only student I would be able to follow in the class. I felt that having more than one focal participant in the classroom would provide a more interesting description of the classroom. Shortly after Ms. Koch agreed to allow me to work with her class, she informed me that she had enlisted Pedro, one of her bright ESL students from her fifth period class, to join second period on the days I visited. While I felt that this was somewhat unusual and rather contrived, I was glad to have a second focal student to work with. I was also happy that Pedro had come so highly recommended by his teacher.

Anna, a fashionable and pretty eleven year old, was recent immigrant from Mexico. Before coming to the U.S. she had lived all over Mexico. She moved to the U.S. two years before, in fourth grade, and attended bilingual classes in fourth and fifth grade. In Mexico, she had attended a public school and studied in Spanish. She wasn't happy with the U.S., because she didn't like the way people dressed, and the fact that young girls had boyfriends, but she did feel that she was getting a better education than she was in Mexico. Although Anna maintained that in Mexico, "school's a mess," she still wanted to return home. "I know it's better opportunities (in the U.S.), but I love Mexico" Anna said.

Speaking, she was much more comfortable in Spanish than in English, but she did seem enthusiastic about using English. At our very first one-on-one interview, Anna asked me if she could "talk Spanish if I don't know something?" I said yes. But, because Anna had a very slight speech impediment—almost a stutter—it was occasionally difficult for me, a nonnative but highly proficient Spanish speaker, to understand her in her native language. Nevertheless, Anna was always eager to repeat or translate her words in English.

Born in the U.S., Pedro, also eleven, told me that he felt more comfortable in Spanish than in English. He always spoke Spanish at home and returned to his grandmother's in Mexico for at least two months in the summer and several weeks during Christmas and spring break. To me, it seemed he was also very comfortable with English—Pedro's English phonology was very native-like. He said that because his friends could "understand (Spanish) but not speak it," he spoke English exclusively with them.

Anna and Pedro sat at a table constructed of five desks, located at the left front of the classroom. Anna and Pedro almost always sat next to each other, with their backs to one wall, facing the rest of the class. They were close to the teacher when she conducted whole class lessons, as well as the board and the overhead projector. In small groups, they were allowed to use Spanish, as the girls who often sat at their table, Sarina, Emily, and sometimes Ginger and/or Kathryn, were familiar with Spanish.

Both Pedro and Anna were good students. Anna said that her grades were mostly 70s and 80s, whereas Pedro said that his grades were usually

80s and 90s. For Pedro, math was his favorite subject because "it's easier to learn." Although she had a sister who was a model, and she herself was quite pretty, Anna was not interested in a future career path like her sister's. Anna was in the U.S. "for better opportunities." She wanted to be a plastic surgeon. As a result, she took school very seriously. In fact, later in the spring semester, when starting a state standardized math practice test in English, Anna burst into tears, because she was so afraid she would do poorly.

Because she had just moved to the U.S. in fourth grade, she was exempt from taking the state standardized tests for English language. She took the tests in Spanish in fifth grade and met minimum standards in Reading and Math, but did not meet the minimum in Science. In fourth grade, Pedro took the state tests in Spanish. His results indicated that he had met minimum expectations in math and writing, but not in reading. In fifth grade he completed the state tests in Spanish in reading and science, but did not meet the minimum expectations required by the state. That year he took the math portion of the state test in English and did meet the minimum.

Although they worked well together in small groups, Anna and Pedro were polar opposites in terms of their class participation in large groups. Anna was very shy and reserved and did not interact in whole class discussions unless she was called upon. She did, however, answer Ms. Koch's questions presented to the whole class. For example, when Ms. Koch would ask (as she regularly did): "What does percent mean?" the whole class would be expected to respond "Over a hundred." In addition to these types of questions, Anna also answered questions as self-talk, in the form of non-interactive speech. In other words, she engaged in the discourse, but with or to no particular interlocutor. She often spoke aloud, but to herself, working out problems orally and answering teacher questions during whole class discussions and when the teacher gave time for students to work individually. In fact, Ms. Koch once confronted Anna about her reticence to speak in class. Anna agreed that she needed to participate more, but admitted that her shyness overcame her.

In contrast with Anna, Pedro was enormously enthusiastic about participating in class discussions and group work. On one February day, Ms. Koch found it necessary to curtail Pedro's participation. She explained that while she was happy he knew the answers, it was important to let others answer questions as well. He acted as though he understood, and gave others a chance to speak. Because he was so eager to answer questions, Pedro was considered very knowledgeable in math by his group members.

Pedro seemed to fit in well in Ms. Koch's second period, despite his just occasional presence. I believe that this resulted from the fact that Pedro was on the same 'team' as the students in second period. The sixth graders at Ritter were on 'teams' in which all of their classes were taught by the same teachers. Therefore, many of the students in Ms. Koch's class knew Pedro from other classes.

In the next chapter, I will paint a picture of how each of these focal students and teachers participated in forming a community of practice in their respective mathematics classes. I will also show how the participants in the three classes worked to create practice that supported or hindered learning of mathematical discourse and concepts for the six focal students that I followed.

NOTES

1. As is common in educational research, names are pseudonyms, and other identifying characteristics have been changed to protect participants.
2. I address each student's fourth and fifth grade basic skills standardized test results. The fourth grade exam tested writing, reading, and mathematics. The fifth grade exam tested reading, mathematics, and science.
3. Mr. Martinez always included this gender-neutral statement when he reminded students of the no-make up rule.
4. Alternative certification participants must have prerequisite undergraduate education. They take specially-designed education classes concurrently when (student) teaching. Alternative certification is awarded upon completion of 1–2 years of approved classroom experience in a public school.
5. Although she explained that "warm-ups" were not allowed at Hot Springs (according to the principal), Ms. Loesely carefully integrated the warm-up activity with the daily lesson.
6. Honors class was first period; the class I followed was not honors and was held third period.
7. In order to maintain the integrity of the participant's meaning and intent, this and all other direct quotations (of all participants) are transcribed exactly as stated.
8. Exemptions from English language standardized tests were usually given only 1 year, and to comply with the No Child Left Behind Act, 2001, exemptions were not given in content areas such as math (content exams were offered in Spanish at SISD), but they were in English language achievement exams (with the exception of ESL placement tests). It was unclear why Nestor was exempt from standardized assessments.

CHAPTER 3

COMMUNITIES OF PRACTICE IN THREE SIXTH-GRADE MATH CLASSES

COMMUNITY OF PRACTICE

The CoP framework (Lave & Wenger, 1991; Wenger, 1998), developed as a learning theory, incorporates the notions of social practice with community, and lays the groundwork for an understanding of the dynamic role of identity in a learning situation. Within the CoP, the socially shared endeavor that is constructed and developed by a community of learners is that which constitutes practice. Wenger (1998) defines a practice as something that "includes . . . language, tools, documents, images, symbols, well-defined roles, specified criteria, codified procedures, regulations, and contracts . . ." (p. 47) The overall social environment, artifacts, and activities in which members participate represent a social practice. In a math classroom, for example, the students construct a discursive social practice with the help of the teacher, curriculum, norms of school, as well as each other.

The CoP framework posits that community can be created in the classroom (or wherever a CoP occurs) when three important criteria are met. These characteristics that comprise the special idea of community (as defined within the framework) are mutual engagement, a joint enterprise such as a task or activity, and a shared repertoire of negotiable resources (Wenger 1998, pp. 72–73). With regard to the first of the three dimensions,

English Language Learners and Math, pages 21–65

Wenger (1998) makes the point that while mutual engagement is a requisite feature in the development of the community, the community "does not entail homogeneity" (p. 75). In the same way that a classroom consists of heterogeneous populations, a CoP represents a diverse community of members. The joint enterprise, or activity that members must carry out refers to the situated interactions between participants engaged in a social practice, and forms a part of the social practice in which all participants involved—both the new, or novice, and the experienced, share. Sometimes the joint enterprise is simply a small task that must be constructed within the social practice and larger, overall activity that subsumes the joint goal or task. Participation implies engagement and individual accountability within a social practice and a group goal or task (Lave & Wenger, 1991). When one does or does not participate in a group goal or task (such as an assignment or duty in class, for example), one's interaction expands or is limited within the particular activity or overall practice. The shared repertoire of negotiable resources includes language and specific ways of using language in certain communities, as well as the tools, artifacts, routines, stories and styles that are used in particular practices (Wenger, 1988).

According to Wenger (1998), these three principal characteristics must be present in order to associate practice with community—which is necessary to establish a particular CoP. It is important to note, though, that a CoP is defined in terms of not just the community membership and the social practice that are collectively and individually carried out, but that the intersection between the two is the critical, defining aspect (Garrett & Baquedano-López, 2002). Both the membership and the practice of a CoP must be examined together to understand how participation (learning) occurs in social settings.

Central to the CoP is the concept of identity. As members progress from 'legitimate peripheral participants' (Lave 1991, p. 68) to full participants in a community, their identity shifts and transforms. The progression from the periphery to full participant status reveals the changing identities of learners as they move along the path of learning. The CoP framework maintains that identity should not be viewed within a dichotomy of social versus individual, but that the interplay of both constructs are vital to the idea of identity. Operationalizing this notion, Wenger (1998) provides a definition of identity as understood in practice. He posits that identity can be characterized as negotiated experience, community membership, learning trajectory, a relation between the local and the global, and as a nexus of multimembership (p. 149). Identity is thus complex and multifaceted and relies heavily on dimensions of practice.

Within the CoP, the process of learning entails identity development and shift. Lave (1991) proposes that "full participation in today's world (is a process of) developing knowledgeably skilled identities" (p. 65). Developing 'knowledgeably skilled' identities is thus of utmost importance in a learning environment. As learners interact within a (learning) community and co-construct knowledge by participating in joint tasks, their social and academic identities both develop and shift to reflect participation in the creation of new knowledge.

A classroom is an example of a heterogeneous community which shares a joint goal, shared repertoire, and mutual engagement in academic and social endeavors. In a classroom, all students are expected to carry out a common, negotiated enterprise, such as the learning of a particular content area subject. The shared repertoire is particularly important in a classroom because it refers not just to common registers and discourse about the content area or topic the students are learning, but also to the shared understandings of the activities and artifacts involved in the overall practice. According to Wenger (1998), "the joint pursuit of an enterprise creates resources for negotiating meaning" (p. 82). He illustrates this point with the explanation that a shared repertoire in a classroom can entail both spelling tests and shooting spitballs (1998, p. 82). Overall, a classroom having all three of these critical dimensions can represent a CoP. Because identity and identity formation are also core issues in the lives of adolescents, the CoP framework is a particularly useful tool for understanding and situating language socialization practices of adolescent language learners in middle school. The analytical focus on identity in the CoP provides a lens for examining issues of identity.

The CoP has as its core unit of analysis the community of practice itself (Wenger, 1998), however, examining individual activities, such as those that emerge from cooperative classroom lessons, can reveal aspects of participant interaction, identity, and evidence of learning target concepts and discourse. An examination of patterns of participation in CoP can reveal how a learner moves from peripheral to full participation in the situated community of learners as members interact within group tasks. As a result, the CoP model can be fruitfully employed to provide an insight into how individual learners of math participate within groups in the practice represented in the reform math class.

Here, I present three communities of practice where student participants' identities shift within their situated practices as they gain knowledge about math language and content. These classroom communities I describe represent social practices which share similarities in their use of mathematics as a practice and as a common 'language' (part of a shared repertoire), but differ in terms of student participation and the degree of success of particular members. In this chapter I will provide a 'thick description' (Geertz, 1973) of each of the three sixth-grade classrooms I worked in at SISD: Ms. Koch's at Ritter Middle School, and Ms. Loesely's and Mr. Martinez' at Hot Springs Middle School. I intend to show how each CoP creates practice that either supports or hinders potential full membership for the focal ELLs and bilinguals in the three mathematics classes.

The classroom participants, including students, teachers, and support personnel, in each of the three sixth-grade classes formed very different, but nonetheless, viable communities of practice in their math classrooms. Each of the three classes shared the three crucial components of a community of practice: joint negotiated enterprise, mutual engagement, and a shared repertoire. In this chapter, I will argue that although each of the three classes represents a community of practice in which learning occurred, the conditions dictated by the teachers, the curriculum, the situated practices, and the intergroup relations played a major role in the efficacy of the learning process. In particular, I will show how issues of participant marginalization, discontinuity in the situated classroom practice, individual identity, and overemphasis on group identity, were present in the three communities. In addition, I show the varied ways in which novices interacted with experts and exhibited agency in their path to learning. These issues differentially affected the focal students in this study.

In examining these three classes, I discovered that the six focal language learners I followed were successful in at least some of their academic endeavors. I define success as success in learning, and this is evidenced in several ways. Evidence of learning is defined as carrying out a requested task successfully, such as answering oral and written content or language-related mathematical problems or questions originating from the book or other 'expert' (like the teacher); responding appropriately to a peer or expert; or teaching content to another. In the three math classes at SISD, some students were able to answer questions posed to them by the teacher, or were able to negotiate a group task presented by the book or the teacher, or were capable of producing correct answers on an assessment or exercise.

THREE COMMUNITIES OF PRACTICE

In defining a community of practice, certain criteria must be met. The community should adhere to the particular dimensions that define a community, and the practice must also be considered an acceptable social practice that entails participation and reification. First of all, the practice implemented in the community should be the primary act of "doing in a historical and social context that… (supplies) structure and meaning" (Wenger, 1998, p. 47). The social practice in math class encompasses the language (math discourse), artifacts (books, pencils, paper, desks, realia), rules and regulations (knowledge of school, how to work in a group), and procedures (how to take a test, write notes, carry out an experiment). Roles and identities also contribute to the shared practice in class. The practice is characterized by shared social participation, be it peripheral or full participation by all members.

Reification of (aspects of) the shared social practice is another core component of a practice (Wenger, 1998). Converting abstractions into reality, or accepting those notions, tools, codes, or symbols as concrete concepts contributes to the establishment of a particular practice. Within a math class informed by reform curricula, the participants of the community of practice must accept the situations, scenarios and tasks as presented by the curriculum, the mathematical problems to solve, and the processes involved in carrying out tasks and problem solving. Wenger (1998) notes that "reification shapes our experience" (p. 59). When students of math understand that they must fulfill the role of landscape architect, baker of brownies for a crowd, expert on cat characteristics, or scratch-off card game player, they agree to accept participant frames that are required of them within the dictated mathematical community of practice of which they are a member.

FOCAL PARTICIPANTS' IDENTITIES AND CLASSROOM ROLES

Teachers

In terms of teaching style, discipline, attitudes toward reform mathematics, and use of CMP, the three classes of Mr. Martinez, Ms. Loesely, and Ms. Koch were rather different. Falling on opposite sides of the spectrum, Mr. Martinez and Ms. Koch represented the poles, while Ms. Loesely fit somewhere near the middle, but resembling more Ms. Koch's style than Mr. Martinez'. Although Ms. Koch's dimensions were somewhat more aligned with the components of CMP, Mr. Martinez' class still worked as a learning community with the goal of achieving mathematical competency. Figure 3.1 outlines how the three teachers could be seen on a continuum with respect to CMP.

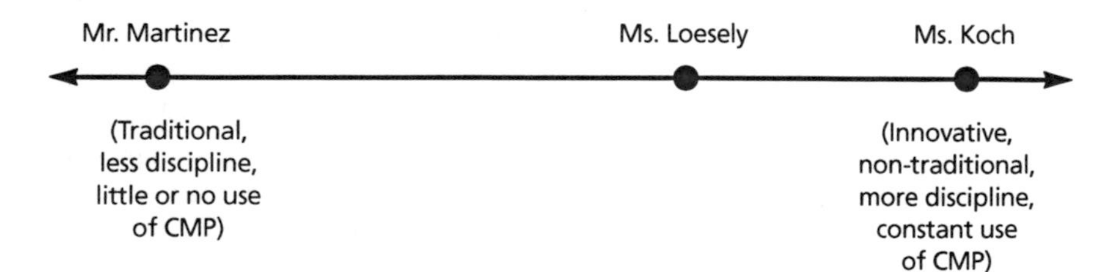

Figure 3.1 Continuum of Teachers—style, discipline, CMP use.

With regard to social and cultural issues, the teachers' identities were also rather different. In terms of their perceived 'character' (whether they were viewed by students as 'mean' or 'nice'), student-held beliefs of each of the teachers were considerably different. Since Mr. Martinez generally allowed students to leave to go to the bathroom or nurse when they wanted, and according to the students, was not a strict teacher, they considered him 'nice'. Ms. Loesely was also considered 'nice', but enforced the school rules more than Mr. Martinez did. Ms. Koch was known as strict and by some, 'mean'. The teachers' representations of their ethnic/linguistic identities were also quite different. Mr. Martinez was the only one of the three who could claim solidarity with the students—in terms of (Latina/o) ethnicity and (Spanish) language use. At times, Ms. Loesely, and to a lesser degree, Ms. Koch, used Spanish as a tool to 'connect' with students, but neither were proficient in Spanish. Mr. Martinez was the only teacher who had intimate knowledge of the neighborhood in which the students lived. He incorporated this knowledge in the classroom to provide references and make connections with his students. Next, I will provide examples how the students and teachers worked to create practice within their communities of practice and expand upon the role that identity played in each of the three math classes.

Mr. Martinez' Class

As mentioned, I consider Mr. Martinez a teacher with a somewhat traditional teaching style and a reduced emphasis on discipline, but a strong relationship with his students. His Latino identity and use of Spanish served to connect to some of Mr. Martinez' students, and in particular, those who were weaker in English than in Spanish. He often called students '*mija*' [*mi hija*, literally 'my daughter', a term of endearment] or *mijo* [*mi hijo*, literally, 'my son']. In addition, his familiarity with the neighborhood in which the students lived also facilitated connections with his students.

Mr. Martinez was not a fan of alternative teaching techniques and rejected the use of realia, hands-on activities, and structured group work, as is encouraged in reform-oriented math curricula such as CMP, and also in programs which aim to serve needs of ELLs such as Sheltered Instruction (Echevarria, Vogt, & Short, 2008; Echevarria & Graves, 2007). In most of Mr. Martinez' classes, he took center stage while directing student focus and interaction to the problems being solved on the overhead projector. Students often knew what to expect from Mr. Martinez' class, but also knew that they could act as agents to change the daily plan if they needed to.

Mr. Martinez was a friend to his students. He joked with them and they joked back. In late spring, the class was working on a state standardized test preparation worksheet which reviewed the concepts they had learned previously in the year. As they are handed the assignment, Mr. Martinez asks the class "What is acute?" to help them review types of angles. A clever and somewhat cheeky bilingual in class named Valerie quickly retorts (in Spanish): "*bonito*!" [meaning good looking or cute]. Not one to miss an opportunity, Mr. Martinez responds with a twinkle in his eye: "No, that's me!" Situations like these were common occurrences in Mr. Martinez' class.

Students were in, out, and generally all over his classroom. Because he ran the after school program, many of his students knew that he could often be found in his classroom and would welcome their visit. During class, he allowed students to come up to his desk freely, and in some cases, allowed students to sit in his desk. With 19 students in the class, there were many to be found out of their seats. Students were usually allowed to leave class to go to the bathroom, get water, or go to the nurse. When selected students were asked to work problems on the overhead projector, Mr. Martinez would allow them to sit at his high stool to write on the transparency. He would get up and stand while they worked the problems.

His background as a carpenter seeped into his daily lessons. Mr. Martinez would often use examples from his days as a contractor to illustrate math problems or potential dangers in getting a math problem wrong. The students were fond of his carpentry anecdotes and liked to comment on how the house they would make would fall down as a consequence of bad mathematical problem solving or decision-making.

Mr. Martinez worked to enable engagement toward a joint negotiated enterprise in his class by drawing all participants together and focusing on him, the overhead projector, and the task at hand. The students' knowledge of the teacher's joking manner, generally lax rules, background as a carpenter, use of Spanish, and knowledge of how the class operated formed part of their repertoire of negotiated resources they shared as a CoP. For example, since Mr. Martinez' students understood that if they yelled out an answer to a question he had posed, the respondent with the loudest voice would have a good chance of being recognized. In the same vein, the students knew that if they asked enough times, that they would eventually be able to leave class to go to the nurse or the bathroom. And although his rules were somewhat relaxed, all of Mr. Martinez' students used the formal title 'sir' to get their teacher's attention. It was unclear to me that the students did this solely in Mr. Martinez' class, or if this was used in other classes with male teachers. It was obvious that while his students felt comfortable with their math teacher, they also respected him as a teacher and a person.

Mr. Martinez' Class as a CoP—January 28

I will draw from a day in Mr. Martinez' class in which CMP was utilized to show how this class as a whole functioned as a CoP. Mr. Martinez' normal routine when using CMP was as follows: he began the lesson with an introductory reading from the text, then conducted a teacher-fronted lesson introducing or reviewing skills in a whole class format on the overhead projector, and then gave the students an individual written assignment to practice the skill they learned. On this early day in the spring semester, Mr. Martinez' class is beginning a unit on fractions. The students have just recently completed a CMP activity where they made fraction strips of various sizes and colors to help them understand and measure different sizes of fractions. Today they have been using the fraction strips they prepared to measure fractions and thereby better understand the concepts presented. In the following example, Mr. Martinez is conducting a whole class lesson on comparing fractions. They have just finished using the manipulatives and are now working out the answers to book problems and inductively coming to a conclusion regarding the strategy they are using to decide which fraction is less or greater than the benchmark fraction. Represented in Example 3.1 are Thomas and Mr. Martinez; Valerie, a loud but bright bilingual; Julie, a bubbly and often off-task English monolingual; and Armando, a bilingual who was good friends with focal students Thomas and Benjamin.

EXAMPLE 3.1
A COMMUNITY OF PRACTICE COMPARING FRACTIONS—MR. MARTINEZ' CLASS, JAN. 28

M = Mr. Martinez
G = Girl
V = Valerie
A = Armando
T = Thomas
B = Boy
J = Julie

1. M: Okay, let's do B.
2. G: Can I do it please, please?
3. V: I wanna do it.
4. T: The five six is higher, *mas que*, more than!
5. A: Sir, does it work the other way around?
6. T: More than!
7. B: More than.
8. M: Okay, Thomas says greater than.
9. G: Aw, that's no fair!
10. A: He said more than, sir!

11. T: Because the six is lower than the eight, and the five is the same!
12. M: Okay.
13. J: I need to go to the restroom!
14. M: Let me explain Thomas' strategy here.
15. A: You said it didn't work the other way around.
16. T: Yeah, but then it's different!
17. M: Something like it, something like it, but no, over here on the first strategy, where the two denominators are the same.
18. M: Cuando los dos denominadores son iguales, comparan los numeradores [when the two denominators are the same, compare the numerators]. When the two denominators are the same, you compare the numerators, when you have numerators the same +/.
19. T: You look at the bottom, which one's less!
20. M: xx numerators, but there's something special about this <you don't say six is less than eight> [>] like they do in elementary school.
21. T: <It's more higher> [<].
22. M: Lemme show you.
23. B: Show us.
24. M: If we went to, Peter Piper Pizza, is that one y'all go to?
25. A: Yeah, on the Parkway.
26. M: If we went there, we ordered a pizza +/.
27. V: Oh, we're going to Peter Piper, I love you!
28. M: No, we're not going, I said if! If we order one that's cut into eight slices +/.
29. M: I put it at this table, and another one over here.
30. A: How come you don't put it on mine?
31. M: Okay, I got there a little bit early, okay, I had one of this, cuz this was pepperoni.
32. G: You eat a lot.
33. M: And I, when I ordered this one, they were both the same sizes, xx.
34. M: Okay, I ate one of these so I have, we have seven eights left, okay? And this one I had cut into twelve, and I ate, one two three four, five of these slices because it was, it was sausage and I like sausage instead of pepperoni.
35. M: I have seven slices, out of eight, and I have seven slices out of twelve.

36.	M:	If I ask you to go help yourself to the pizza, which one would you head to? One slice only, one slice only!
37.	T:	Seven eighths!
38.	A:	Seven, seven eights.
39.	M:	Why?
40.	T:	Cuz there more left!
41.	M:	Wait, hold it, listen, listen. I hope the answer's not because he likes pepperoni.
42.	M:	Why?
43.	A:	Cuz it's bigger.
44.	M:	What's bigger?
45.	T:	It has more!
46.	M:	So, the smaller the number, the bigger the slice, the bigger the number, the smaller the slice.

In this example from January 28, we see that this busy class, characterized by a good deal of overlapping speech, operates as a cohesive CoP, despite the fact that so much activity is occurring at once. Many people are speaking at once, some students are up walking around the classroom, another is sharpening her pencil, one is asking to go to the bathroom and ignoring the whole class' focus on the fractions, while others are taking the pizza discussion out of the mathematical context that the teacher intended. Nevertheless, Mr. Martinez, along with several of the particularly on-task students ultimately directs the class's discourse to achieve an academic goal: to understand how to compare fractions when denominators or numerators are the same.

The progression moves from first examining problem B, where both fractions to compare had the same denominator, then follows an indirect but connected path. Next, we observe on and off-task input regarding one's need to go to the restroom (line 13), Armando questioning the process of comparing fractions with the same denominator (lines 5, 15), complaints that the teacher gave too much weight to Thomas' response (line 9) (which, according to Armando, was not accurately reported) (line 10), then discussions centering around pizza (lines 24–29). The final, crucial mathematical point of the lesson emerges in lines 37–46: the smaller the denominator, the bigger the fraction, and the bigger the denominator, the smaller the fraction.

In accordance with the tripartite criteria of a community of practice, Mr. Martinez' class was mutually engaged in a joint negotiated enterprise—in the form of whole class academic discussion and the comparison of fractions. Despite the outward appearance of disorganization, most of the students were following along in the book and also with the teacher's guidance to compare the fractions in the text to the benchmarks with which the fractions were associated. Since whole class lessons were commonplace with Mr. Martinez, his students understood the rules of engagement of such a lesson. For example, they knew that to gain the attention of the teacher, they simply needed to let him know—loudly, and repeatedly—that they knew the answer and/or that they wanted to participate (lines 2, 3, 4, 6, 7, for example).

The repertoire shared by Mr. Martinez' class included the understanding of effective engagement, as well as rules of interaction. The students knew that they could joke and play with Mr. Martinez, as he did with them. In lines 27, 30, and 32, the students tease their teacher about pizza, and he allows them to do so. Similarly, the students accept that in this class, it is not uncommon for their teacher to elucidate particular mathematical points by using examples such as going to the pizza parlor. He draws in the students to this example by focusing their attention to the pizza restaurant in their neighborhood—the one he knows they likely go to. Since he also lives in the neighborhood, he can make this example particularly relevant for them by describing "their" pizza parlor—Peter Piper Pizza. These examples, or frames, have become an integral part of the shared repertoire of this, and of other math classes which use CMP.

Finally, it is interesting, but not at all unusual in this class for Mr. Martinez to draw attention to Thomas' contributions to the class. In lines 8, 14, 39, 41, and 46, the teacher makes explicit reference to Thomas' comments, questions, or input, and directs the class to pay attention to Thomas as well. Mr. Martinez' recognition of Thomas' contributions accords Thomas high status. As a novice, Thomas exhibits a great deal of agency in the classroom. Often, as in lines 8 and 14 of Example 3.1, Mr. Martinez credits Thomas' contributions, thereby affording them an even higher status as contributions of an expert. Student reactions to Thomas are varied. Today the class has generally tolerated the attention to Thomas, but on other days, some students make comments about him as a teacher's pet, or 'know it all'. Many do not criticize him, though, as he is a useful resource when they don't understand the lesson and are in need of peer tutoring. This example from January 28 reflects Thomas' discursive academic input as an integral part of Mr. Martinez' second period class's shared repertoire.

Thomas

When I asked Thomas what his favorite class in school was, he said P.E. and math. He explained: "Cuz I like P.E. cuz it has sports where you have fun, right? And math cuz I'm real good and used to it, and I know it a lot, and sometimes I learn something new about it, and it makes me more, more better at it." For Thomas, math was an exciting challenge. He knew he was the star student in the class because Mr. Martinez always called on him by name to answer questions and often directed the class' attention to

Thomas and his contributions. For example, Mr. Martinez would say things like "Everybody listen to Thomas" or "*¿Qué (estrategia) está usando Thomas para figurar la respuesta aquí?* [What (strategy) is Thomas using to figure out the answer here?]" Mr. Martinez also often praised Thomas by name in front of the other students. The teacher would jokingly make comments to other students such as "Thomas is gonna beat you . . . (because) Thomas is smart" or "Thomas knows everything." Once, on February 11, when Mr. Martinez recognized Thomas for knowing the least common denominator of a set of fractions, other students voiced their complaints. Example 3.2 illustrates how Thomas was viewed in class. David is a heavyset, comical kid who participated regularly in class discussions. Kelly is a talkative and pretty girl who asked to go to the bathroom in class almost daily.

EXAMPLE 3.2
THOMAS IS ALWAYS CORRECT, FEBRUARY 11

M = Mr. Martinez
T = Thomas
D = David
K = Kelly

1. M: Twenty-three, one and eight hundred seventy-five thousands, is equal to one, and eight hundred and seventy-five thousands, what's the least common denominator?
2. T: A hundred and twenty-five.
3. M: A hundred and twenty-five, Thomas is correct.
4. D: Thomas is always correct!
5. T: He's always correct because he's paying attention.
6. D: It gets kinda old right?
7. K: Yeah.

Although he was often openly chastised and teased for getting the correct answers to Mr. Martinez' questions, other students still relied on Thomas for help on their assignments. On January 28, after Mr. Martinez assigned the class a task to complete individually, several students—including David—could be heard saying: "Thomas, come here!"

Thomas was often willing to help out others, but not until he finished his own work first. On February 4, he became frustrated when he couldn't get his own work done before having to help others. "I woulda been finished already but it weren't for someone" he said, implying that his neighbors took too much of his time. According to Armando, Thomas' friend and neighbor in class "(Thomas') way of helping is just finishing his work and then helping." But Thomas felt differently. When asked about whom he worked with in groups, he explained that he didn't really *work* in groups, "I just help people." Thomas was often regarded an expert by both his teacher and other students—despite the fact that he was a novice learning language, math discourse, and new mathematical concepts in sixth grade.

For Thomas, math class was a place to be successful and have fun in school. He was good at math and worked at being the best in class. He couldn't understand why Mr. Martinez gave him progress reports because he always did so well. When handed a progress report to get signed and return, Thomas asked "Is this really necessary, I always get a ninety-eight on my progress reports." Sitting up front, right by Mr. Martinez, Thomas was almost always engaged in class discussions. On those days in which Mr. Martinez didn't recognize his efforts as Thomas felt the teacher should have, he would make himself known. Example 3.3 from February 4th reveals Thomas' identity as the best student in class and situates his identity in the classroom CoP. It further illustrates how Thomas participates in whole class discussions and maintains his academic identity. In the lesson, the class has been enlisted in the task of comparing two fractions. They must determine if the fractions are equal or if one is greater or lesser than the other.

EXAMPLE 3.3
THOMAS, CALM DOWN! FEBRUARY 4

M = Mr. Martinez
T = Thomas
A = Armando
K = Kelly
AL = All
V = Valerie

1. M: Number twelve.
2. T: Equals.

———

3. T: They're equal, sir.
   (students are getting noisy)
4. T: No, they're equal.
5. A: They're equal sir, cuz all you needa do is drop off the zero.
6. T: Yeah, they're equal sir.
7. T: Just put on the equals sign!
8. T: Just put on three zeros on those!
   (Thomas is yelling and is very enthusiastic)

9. K: Thomas, calm down!
10. M: You're telling me that one is equal to a thousand?
11. AL: Yes!
12. A: Cuz all you need to do is match up the zeros!
13. V: Yeah, that's equal to.
14. T: Or add the zeros.
15. M: Man, you're, I can't fool you kids.
16. V: No you can't.

On this particular day, Thomas had to vie for the teacher's attention a bit harder than he normally had to. As usual, he was on task and answering the teacher's questions appropriately, but because other classmates were also providing thoughtful and in some cases, correct responses to the teacher's queries, Thomas was motivated to be recognized. In lines 2, 3, and 4, Thomas attempts to engage the teacher. When his friend Armando supplies a rationale as to why the two fractions they are comparing are equal (line 5), Thomas jumps on the bandwagon and agrees with Armando (line 6). After that, though, Thomas attempts to capture Mr. Martinez' attention with enthusiasm and volume (lines 7, 8). As he does this, though, he is chastised by Kelly, a loudmouthed girl who often relied on Thomas' assistance on assignments. Shortly thereafter, Thomas hangs back, and the teacher confirms that 1000 is equal to one (line 10). Part of their shared repertoire, Mr. Martinez' joking manner draws in the classroom participants so that they mutually (lines 11–14) and successfully engage in the fraction comparison activity (their joint negotiated goal). Next, we will see how Benjamin engages in the classroom CoP.

Benjamin

Unlike Thomas, Benjamin did not have strong feelings about math. His favorite class in school was science, because as he said "I don't know that much words, like, I want know more (science words)." He also liked that they had art and P.E. in American schools. But for Benjamin, math was just okay. It was quite different from the way math was taught in Mexico, according to Benjamin. Since he had just moved to the U.S. only a year and a half earlier, his views on math were influenced by his experience in Mexico. For Benjamin, math class was quite different in the U.S. than in Mexico. He explained:

> Like, um, they put us, like, put us to do it, and they just put it on the, the, problems on the board and then we have to do it, they don't do it, and they ask us if we're done, if we say yes, they do on the board...

In fact, contrary to popular belief, Benjamin believed that the math concepts that were taught in the U.S. were more difficult than in Mexico.

Benjamin was not engaged in the math lessons in the same way that Thomas was. Benjamin felt comfortable participating in whole class group discussions, but was not generally successful in terms of answering questions correctly, or getting the right answers. When the teacher asked the class a question that Benjamin didn't know, Benjamin responded "I dunno," often to no one in particular. A typical example is seen on February 11, when Mr. Martinez posed the question: "How many fours are in twelve?" Several students, including Benjamin responded "Three." But when Mr. Martinez corrects himself and rephrases the query to "I mean in fourteen," Benjamin replies, "Uh, I dunno."

Mr. Martinez called on Benjamin every once in a while, but Benjamin's stock answer was usually "I dunno." He did not feel comfortable answering questions in class individually, and especially in Spanish. Benjamin preferred to use English in school, but Mr. Martinez sometimes posed questions to Benjamin in his home language, Spanish, as in Example 3.4. The teacher had just asked the class if six was a factor of one hundred. When they say no as a whole class, he asks if seven is a factor of one hundred. They start to yell out 'no' responses.

EXAMPLE 3.4
BENJAMIN: I DUNNO, FEBRUARY 11

M = Mr. Martinez
AL = All
B = Benjamin
G = Girl

1. M: *Siete, siete, por qué, es un cien* [Seven, seven, why, it's a hundred].
2. AL: No, no!
3. M: *Benjamin, siete, por qué* [Benjamin, seven, why]?
4. M: Seven times what?
5. B: [= coughs] ## I dunno sir.
6. M: There is a seven, look, one, two three four five six seven!
7. G: No!
8. B: But it doesn't go to a hundred!

The teacher was trying to understand why (and if) Benjamin believed that seven was a factor of a hundred (lines 1, 3). When he confronts Benjamin to ask him why seven could have been a factor, Benjamin becomes nervous and retreats with his stock answer, "I dunno" (line 5). Nevertheless, Benjamin indicates in line 8 that he understood the concept, by explaining that seven doesn't go to a hundred. His use of 'I dunno' in this case reflects Benjamin's insecurity and self-doubt in math class. His participation in class mirrored his identity as somewhat social and good-natured, but in terms of math, his capability was average, and he lacked confidence.

Benjamin was not totally invisible in math class; he was just relatively quiet and generally went unnoticed, except by his friends. He was by no means viewed by the other kids as a whiz at math, but then again, he wasn't seen as one of the 'dumb kids'. Besides Thomas and Armando, his bilingual neighbors, Benjamin did not engage with many other individuals in class. He knew many in class, including the newcomer sisters from Mexico from his ESL class, but did not speak to them in math. A good-humored kid, he laughed at others' jokes and kept up with the social goings-on in class. If for example, girls in the back of the room were reprimanded for putting on makeup in class, Benjamin would track the situation. He also followed the teacher's direction and instructions on academic tasks. If necessary, he would occasionally elicit help from Thomas or Armando, but usually tried to complete the work himself.

In Example 3.5 we see Benjamin participating actively within a whole class setting. The task is to determine the greatest common factors (GCF) of 1000 and of 448. As the example begins, the students are looking for the GCF of 1000.

EXAMPLE 3.5
GCF: BENJAMIN PARTICIPATING SUCCESSFULLY AND UNSUCCESSFULLY IN HIS COP, FEBRUARY 11

M = Mr. Martinez
B = Benjamin
V = Valerie
D = David
T = Thomas
G = Girl
A = Armando

1. M: What do you divide it by?
2. B: By four, by four!
3. V: By four!
4. D: Told you.
5. B: Four!

6. M: Four?
7. B: Yes, divide it by four.
8. G: You can divide it by four, but can you divide it by more?

———

(They decide that 8 is a larger factor of 1000 and 448)

9. M: We found out we can divide a thousand by eight, can we divide four hundred and forty-eight by eight?
10. G: No.
11. B: Yes.
12. A: Yes.
13. T: Because forty-eight, it's um, it's um eight.
14. M: *Ocho en cuatrocientos cuarenta y ocho, ¿Sí o no?* [Eight in four hundred forty-eight, yes or no?]
15. B: Yeees!
16. M: *Sí* [yes].

———

17. M: It was such a big number we had to go back to use factor trees cuz I didn't know whether we had reduced them or not, okay, somebody find all the common factors.
18. G: Two!
19. A: Two to the power of +/.
20. B: [= laughs].
21. M: Okay so what is two times two times two?
22. M: So we divided up here by eight, so are we in lowest terms?
23. B: Yes!
24. M: Well, we have to make sure, I didn't know if I was in lowest terms.
25. D: Well you should!
26. B: [=! laughs].
27. M: If I did, I wasn't gonna tell you.
28. B: [=! laughs].
29. M: Okay so what was the prime factorization of a thousand?
30. M: Who remembers?
31. B: Eight, It was eight!
32. M: Prime factorization of a thousand, what was it?
33. B: Eight!
34. G: Uh, two times two.
35. B: I dunno.
36. M: Okay so the prime factorization of a thousand is two times two times two times five times five times five, is that correct, is this correct?

37. B: [=! laughs].

38. B: Yes.
39. M: Okay so we're gonna do a thousand all the time.

This example illustrates not only Benjamin's participation in whole class academic endeavors, but also how the class interacts and engages on a daily basis as a CoP. At the beginning of the example, the students spend time determining what the GCF of 1000 and 448 is. Lines 1–8 track the class participants' negotiation of the answer. Here Benjamin is enthusiastically engaged in this group negotiation. Benjamin's responses to the teacher's question (lines 2, 5, and 7) are correct, but he has not supplied what the teacher is actually looking for, the GCF. As the discussion continues, he continues to be engaged in the next task—that of determining the GCF of 448 (lines 9–16). When Thomas suggests in line 13 that eight is the GCF, Mr. Martinez confirms Thomas's hypothesis in Spanish (line 14). Benjamin then wholeheartedly agrees with this answer (line 15). Benjamin's answer is correct. A few seconds later, Mr. Martinez reintroduces the idea of factor trees and common factors (line 17). The students are familiar with this concept, and are also familiar with 'prime factorization' (discussed in lines 29, 32, 36) but Benjamin's responses in lines 31 and 33 indicate that he does not actually have the answer the teacher is looking for. After he has been cued by a female student that 'prime factorization' is different from 'GCF' Benjamin falls back on his stock answer "I dunno" and briefly retreats from participation. Shortly afterwards, though, Benjamin is pulled back into the lesson and re-engages in the discourse (lines 37, 38).

In this example from February 11, Benjamin was an active and engaged participant in much of the classroom discourse. His success was somewhat less consistent than his participation (in that he did not come up with the 'greatest' common factor), but he nevertheless exhibited a shared understanding of how to participate in the CoP. In general, all of the classroom students were following the teacher's questioning. They were thus mutually engaged in the task of determining GCFs and prime factorizations—their joint negotiated goal. The repertoire shared by Mr. Martinez' class is evident in the example from February 11.

The shared repertoire of this community is illustrated in several instances in Example 3.5. The first is exemplified in lines 2–8, as the students provide rapid-fire responses to the teacher's questions. The students (Benjamin included) understand that calling out (without being acknowledged) is appropriate, and indeed, effective if one's answer is to be acknowledged. Mr. Martinez' (and others') use of Spanish also contributes to the shared repertoire. And joking with the teacher and students (lines 25, 27) is a part of the shared repertoire. Finally, the use of mathematical terminology and discourse is undoubtedly shared by the CoP in math class. Mr. Martinez introduces a number of important features of mathematical discourse that are familiar to the students, but not always crystal clear in terms of meaning. In this example, the teacher discusses GCF, factor trees, prime factorization, and common factors. But certain instances of student discourse show that some learners, including Benjamin, do not fully comprehend all of the mathematical concepts that the teacher presents. In lines 18 and 19 Armando makes a joke about exponential numbers that Benjamin thinks is funny. It is not actually clear if Benjamin has understood this referent, though.

In the same way that Benjamin misunderstood Mr. Martinez' question about prime factorization (lines 29–36), Armando's math joke shows that he and others have a familiarity with features of math discourse, but may not fully understand the mathematical meaning undergirding the concept. However, it is clear that this classroom represents a coherent CoP in which learning occurs for some participants. Thomas represents a successful legitimate participant, while Benjamin is also a legitimate participant moving from much farther from the periphery to a potential full participant. Although Benjamin was not generally successful in class, he was progressing toward mathematical discourse development and success in math. As the semester ensued, Benjamin participated more and in more appropriate ways. Examples 3.4 and 3.5 show Benjamin participating on the periphery. He has learned some ways of interacting in math, but has more to go.

In the next segment of this chapter, we will visit Ms. Loesely's class and examine how American-born Jennifer and Nestor participated in their classroom community of practice.

Ms. Loesely's Class

At first glance, Ms. Loesely's class was more organized, disciplined, and cohesive (in terms of numbers of students working cooperatively and on-task) than Mr. Martinez' class. Very often Ms. Loesely's class worked in groups and used the open-ended small group discussion format, along with the requisite realia advocated in the CMP curriculum. She was an eclectic teacher, incorporating varied teaching styles in her lessons, and often tailored her teaching method to fit with the day's lessons. When it was necessary for students to work in small groups, she usually had the tables prearranged to make tables of 4–5 students. When she planned to start the lessons with a whole class discussion, review, or other activity and then follow with an activity requiring group work, Ms. Loesely would explain the instructions for the lesson and then allow a particular amount of time—

often no more than 30 slow seconds, counted out loud by the teacher herself, for the students to arrange the classroom furniture appropriate to the day's activity. Since they were accustomed to changes such as desk moving and materials gathering, Ms. Loesely's third period students could carry out classroom directives without an inordinate amount of noise or chaos.

Ms. Loesely had been a teacher of CMP for several years, and a math teacher for many more. Therefore, she was quite familiar with manipulatives for math. She had boxes and boxes full of dominoes, blocks, balls, wikki sticks and the like stashed away in sandwich bags for eventual student use. She explained to me that her garage at home was also filled with math toys, games, and manipulatives from years past. When I inquired as to why she had so many of these things, Ms. Loesely informed me that although CMP had not been around for too long, it was based on several earlier reform-influenced curricula[1] that she had used before. She was, therefore, comfortable with the realia that were to be used with CMP and did in fact use them.

Ms. Loesely's relationship with her students was solid. She was not a strong disciplinarian, but the small class size of 17 students in third period contributed to few discipline problems. Ms. Loesely's discussions of her own cats held the interest of her students, as did discussions of the class mascot: Bill. Bill was a Japanese beta fighter fish who lived in Ms. Loesely's class during the first semester of the school year. When the teacher had taken Bill home for the winter holiday, she made the mistake of accidently pouring Bill in her coffee cup, instead of his bowl, when she was cleaning it. Her presentation of this misfortune delighted, and at the same time made many of her students somewhat uneasy. As a result, she bought a second fish, predictably named Bill 2. Bill 2 was a common topic of conversation in class. He did not often—if ever—serve to direct classroom discourse toward more academic topics, but his presence did help create the classroom as a cohesive community and a stable CoP by contributing to a shared repertoire.

Ms. Loesely also created cohesion and community in her class by repeating important academic and mathematical rules and concepts over and over. She did this so often in fact that some of her students would parrot her sentences. With regard to taking notes, she always stressed: "Anything I write on the board, write on your paper." When explaining how to divide fractions she would say, "The top number goes in the division house." The students would often chime in on these directives and descriptions, repeating them in a mocking, but light tone.

To promote focus and mutual engagement on the group's joint enterprise, Ms. Loesely always listed the day's goals and objectives on the board, so the class knew what concept and task they would tackle that particular day. In addition to the daily agenda, she usually began lessons with some kind of mini-lesson/warm-up to introduce students to the new concepts.

Often, but not always, the students were allowed to work together to find answers to these preliminary activities.

Ms. Loesely's Class as a CoP, January 21

In the following examples from the beginning of the spring term, I will show how Ms. Loesely's third period functioned as a classroom CoP in a whole-class setting. School had just recently recommenced from the holiday vacation, and Ms. Loesely had spent the first week or so back from the holidays wrapping up lessons started in December. On January 21, third period had just finished making their fraction kits[2] the day before (as the other sixth grade math students in Mr. Martinez' class would soon do). On this day, Ms. Loesely had planned an activity for the students to complete using their fraction kits. The task is to complete a preliminary activity in small groups to which she assigns the students as they walk in. In addition to Ms. Loesely and Nestor, other students represented in the following example include Kristina, a bilingual who was somewhat good at math, Jesus, an outgoing ELL who sported new expensive basketball clothes and was very interested in class gossip, and Noel, an English speaker who liked math.

EXAMPLE 3.6
LOESELY, FRACTION WARM-UP, JANUARY 21

L = Ms. Loesely
K = Kristina
J = Jesus
N = Nestor
NO = Noel

1. L: Draw a number and go find your group and sit down, wait, draw a number.
2. L: I didn't say continue talking, sit down and get to work.
3. K: xxx sit down?
4. L: No, you have a warm-up, everybody.
5. J: Hey ya'll, help me.
6. L: You people have to xx.
7. J: What do we have to do?
8. N: I dunno.

—

9. L: You have to answer roll.
10. L: Kristina, Tessy, Sally, Yolanda.
11. NO: Arthur is not here, he got suspended.
12. L: Stephen, Arthur, Noel.

13. J: Who was he fighting?
14. NO: Miss, Arthur's not here, I think, he was in a fight yesterday.
15. L: He was in a fight yesterday?
16. L: Okay, Jon, Claudia, Ashley, Tiffany.
17. J: Oh yeah, Devin started kicking him, right?
18. N: Yeah.
19. L: Nestor, Jesus, I saw Jesus already and Jennifer.

———

20. L: Okay, you have about five or ten minutes to get this problem solved, so get busy.

Example 3.6 appears to show a regular classroom generally off-task and preparing for their daily class. But on closer inspection, this example reveals the inner workings of a group of community participants who share a good deal of knowledge and resources to be able to work as a cohesive and coherent whole. The teacher's first utterances are directives to the students to start to work. It appears that they have trouble doing so, due to lack of understanding of the task. However, they begin to ask questions to clarify their confusion. Lines 3, 5, and 7 indicate that Kristina and Jesus are motivated to engage in the group activity, but they seek further information. Relatively quiet Nestor responds to their queries, but indicates he does not know what they need to be doing (line 8). When Ms. Loesely explains that the students need to both attend to their warm-up and answer the roll call, most of the class members apply themselves to these tasks. The group negotiated enterprise that is dictated by the teacher is carried out once the classroom participants gain an understanding of what they are to be doing. But while roll is called, several students, namely Noel and Jesus, become engaged in a social discourse, which is not wholly unrelated to one of the math teacher's directives. Noel, Jesus, and Nestor succeed in engaging the teacher, and many other silent participants in the discussion regarding the fight that Arthur, a relatively troublesome student in their class, was involved in (lines 11–18). It is clear that a number of class members share knowledge about this fight, in much the same way that they identify with an understanding of answering roll call, drawing a number to be grouped for an activity, or doing out a warm-up at the beginning of class. It just happens that discussing a fight is much more interesting. I argue that the introduction of this fight into the classroom discourse is part and parcel of the shared repertoire, joint enterprise, and also the mutual engagement of Ms. Loesely's third period classroom CoP. Since Arthur was a member of the class, it was the business of all the students in class, as well as the teacher, to be 'in the know' with regard to the social on-goings of the school. What is more, discussion of the fight served as a bridge from the introductory phase of the class' daily lesson to the content that they would later cover.

Focal students Nestor and Jennifer were often engaged in the joint negotiated enterprise in Ms. Loesely's class, but their levels of participation and success in mathematical tasks were often inconsistent. Neither Nestor nor Jennifer was often successful in math in terms of grades, standardized test scores, or teacher-made tests, but both of these students contributed to the CoP that developed in Ms. Loesely's third period. In the section that follows, I present a description of how both Jennifer and Nestor saw their math class, as well as how they were seen by others.

Jennifer

Jennifer liked math class because she enjoyed being with her friends. Many of her bilingual girlfriends were in third period math with her, and this made it pleasant. Jennifer almost always worked with her bilingual girlfriends in class. She was particularly fond of working with Kristina because they liked to talk about social matters. "Sometimes we just wanna be like her and me, like that, until Ms. Loesely comes to check our work." Jennifer explained that when she and Kristina worked together, they also got their assignments done, because Kristina was fast at completing math work. They usually used English in class, but Jennifer said that whenever she and her girlfriends "start fighting, we start talking in Spanish."

Her favorite subject was not math, but English, because she liked to read. Since her older eighth grade sister was good at math but bad at reading, Jennifer could help her sister in reading and her sister could help her in math. Although she was just recently exited from ESL support classes, Jennifer did not struggle to find words in English. She was well-spoken, polite, and exuded an air of maturity. She used polite discourse at appropriate times (please, thank you, ma'am), did not curse in front of adults (as many of her classmates did), and was generally very well behaved. At 14 years old, she was also at least 2 years older than her classmates. Most of the other sixth graders were either 11 or 12 years old. Most likely she was held back and/or started school late, but I never learned the true reason she started middle school at 14.

Jennifer was often quite participatory and demonstrated her ability to think out of the box while using mathematical language when she had an interesting, motivating, thought-provoking task. She was not as proficient in getting "correct" answers as some of the more successful students in class, but she was able to think about and discuss the process involved in finding solutions, as she did on February 2. On this day, the students were learning how to double and triple recipes using whole numbers and fractions. A student has just read this introduction from the CMP *Bits and Pieces* textbook:

> Next week the eighth graders from Sturgis Middle school are attending school camp, Samantha, Michael, Romeo, and Harold have the job of making brownies for an afternoon snack for the entire camp, all two hundred and forty people! (Lappan et al., 2002a, p. 32)

The questions the students were to discuss were the following:

Suppose you get to decide which size brownies will be served to the campers. Tell which size you would choose in each situation below. Explain your answer.

1. You are in charge of buying the ingredients, and you have a limited budget.
2. You have to help make the brownies.
3. You don't have to do any work, you just get to eat the brownies. (Lappan et al., 2002a, p. 33).

Ms. Loesely first conducts a whole class discussion to determine how many batches of brownies are enough, and then instructs the students to answer the questions from the book. Jennifer is working with her bilingual friend, Kristina, but Kristina has walked off at first, then returns and listens to Jennifer. Example 3.7 shows Jennifer working through the problem.

EXAMPLE 3.7
JENNIFER THINKS OUT OF THE (BROWNIE) BOX, FEBRUARY 2

L = Ms. Loesely
J = Jennifer
A = All
B = Boy
G = Girl

1. L: So if we make one batch of brownies are we gonna have enough?
2. J: No!
3. A: No!
4. L: So what are we gonna need to do?
5. B: Cut 'em small!
6. J: Or you can just make more than one.
7. L: Huh?
8. G: Make more than one!
9. L: Make more than one batch!

10. L: Well how are we gonna find out how many we need, talk it over.
11. L: How can we figure out how many batches of brownies we need?
12. L: Turn around and talk to your people around you. (Kristina returns and listens to her partner)
13. J: They're gonna have to make like five batches, cuz # wait, you're not just gonna sell two hundred forty things uh, to people in tiny pieces of brownies.
14. J: Uh, I know, that if you, you're gonna have to make more than one batch because, not unless your gonna put, not unless make em thirty small brownies for each pan # I think that.

———

15. L: What, we can only make, we can only make the recipe stipulates we can only make fifteen large, twenty medium and thirty small, we've gotta be bound by those pan sizes, what are we gonna do?

———

16. J: And then what if you go over the thing and you need more money, so, if like you make the large ones, you'll have to, if you buy all ingredients and you still need more, you have to put some of your own money in, and the second one, you'll probably just like, well, you probably wouldn't want to make the brownies, cuz, if you have to buy all the things.
17. L: How would you answer number one, have you two decided how you'll answer number one?
18. J: If um, if you put the big one, if you make the fifteen, what if you go over your budget, and you have to put some more money from your pocket?
19. L: Yeah, you'd put from your pocket and you wouldn't be able to make enough for everybody, yeah, good thought.
20. J: It wouldn't be fair if you just didn't do any work and all you had to do was eat the brownies.

Here, Jennifer portrays herself as both a competent, rational thinker and an arbiter of fairness. Her mature character reflected concern that the one who was responsible for making the brownies might have to either spend one's own money or do work that others wouldn't have to do—if the bigger brownies were chosen. Because she has taken the role of leader in the discursive interaction, she presumes some expertise in the matter of brownie sizes.

When Ms. Loesely presented the problem to the students and asked if one batch would be enough (line 1), Jennifer enthusiastically engaged in the discourse and suggested a solution (lines 2, 6). Although the teacher did not understand Jennifer's solution (line 7), she had in fact supplied the response the teacher wanted. When Jennifer and her partner, Kristina, begin to discuss the questions, Jennifer becomes quite enthusiastic in evaluating the brownie problem and employs creative, higher order thinking to tackle the questions. Jennifer thinks creatively, and analyzes, assesses, and evaluates the questions that the text and her teacher presents. Higher-order thinking skills such as these are important in open-ended questioning in CMP. In lines 13–14, Jennifer maintains that tiny pieces of brownie would not work to serve 240 students, and suggests that 5 batches might be more effective. She then reiterates the need to make several batches of brownies, citing the inefficacy of serving very small servings to students. Then, Jennifer addresses questions one and two and explains how having to make more brownies might potentially cause the brownie maker to spend their own money to feed their classmates (lines 16, 18). She addresses question 3 in line 20, maintaining that it just wouldn't be fair to make the brownies when others weren't responsible for anything.

On this day, Jennifer demonstrated creativity and skill in using higher order thinking. There was not a right or wrong answer to the lesson, and she was able to think through her solutions to the problems posed in the CMP text. But on other days, there was a right or wrong answer she was expected to reach.

Example 3.8 reveals Jennifer's identity when reaching correct answers in a partner situation. Yesterday the class completed the activity where they determined the size of brownies to make, and answered the three questions regarding the problem of making brownies for 240 people. Since they decided to make the large brownies, and they need to have 240 brownies, the students' task for today, February 3, is to increase the recipe from one pan to make 16 pans of brownies. Jennifer is working with Kristina, her favorite math partner.

EXAMPLE 3.8
CALCULATING INGREDIENTS FOR 16 PANS OF BROWNIES, FEBRUARY 3

L = Ms. Loesely
J = Jennifer
K = Kristina

1. J: Okay, one fourth is twenty-five, right?

2. J: So, point twenty-five, times that by sixteen equals point, four point, there's
3. J: what?
4. K: Four point?
5. J: No, don't look at it, put point twenty five times sixteen.

—

6. J: Oh look I got it, four point; four point, four point, miss! (Teacher comes over)
7. M: Which one are you working on, the sugar?
8. J: Uhhuh.
9. M: That's correct!
10. J: Four point.
11. M: Point twenty-five times sixteen, right?
12. K: So we put four?
13. L: Four cups, uhhuh.
14. J: Okay, we have one half, yeah, fifty.
15. K: Fifty what, cups cold butter?
16. J: Uhhuh.

—

17. J: One egg, so, that'll be sixteen eggs, of course!
18. K: One, oh, one times sixteen equals sixteen each, duh!
19. J: So we're gonna need sixteen teaspoons (of vanilla, the next ingredient).

—

20. K: One half, it's fifty right?
21. J: Uhhuh, fifty teaspoons of baking powder.
22. K: That's lots of fat, isn't it?
23. J: Yes, that is a lot.
24. J: No, baking powder, it doesn't do anything to you.
25. K: No?
26. J: (reading) Cups of cold margarine, this one, does something to you.
27. K: That's a lot of butter.

—

28. K: And three-fourths cup chopped nuts.
29. J: It'll be, seventy, point seventy-five?
30. K: Three, eh, my hair.
31. J: I put point seventy-five?
32. K: Point seventy-five times sixteen equals?
33. J: Twelve.
34. K: Twelve.
35. J: Miss we're done!

The original recipe called for one-fourth of a cup sugar, 1 egg, 1 teaspoon vanilla, one half of a cup cold butter or margarine, and one half baking powder, in addition to flour, cocoa, chocolate, and nuts. Working as a team, Jennifer and Kristina were able to complete the task of multiplying the recipe by sixteen. Jennifer was comfortable using learning strategies of confirming and clarifying the task with her partner (lines 1, 2, 31) as well was Kristina with Jennifer. As they weaved in and out of academic discussion, they succeeded in increasing the recipe to yield 240 brownies. However, they weren't always successful in attaining the 'correct' answers. In the previous example, Jennifer and her friend successfully increased the yield of sugar (lines 1–13), eggs (lines 17–18), and vanilla (line 19). But they did not come up with the correct measurements for butter/margarine (lines 14–16) or baking powder (lines 20–24).

As Jennifer had mentioned, she was quite fond of Kristina and enjoyed working with her in math class. They were used to working together, and their interactions corroborate this. In the same way that the whole class shared knowledge of the shared repertoire of brownie making, these two bilingual friends also shared a repertoire. Their encouraging discourse helped the other to reach the answer they ultimately would decide on. In line 5, Jennifer reprimands Kristina with "Don't look at it" and encourages her friend to focus on the task. Line 17 illustrates Jennifer's sudden awareness that 1 egg multiplied by 16 would equal 16. Her use of 'of course' motivates Kristina to that same realization. Kristina's 'duh' (line 18) indicates that she has also understood the egg calculation. Then, in line 19, Jennifer transfers her knowledge of the egg calculation to the vanilla measurement. The shared repertoire extends also to the calculations they did not carry out correctly. In line 20, Kristina asks Jennifer, 'It's 50, right?' to which her partner agrees.

Despite their inconsistency in attaining the correct calculations, the girls did share a repertoire that aided them in working cooperatively. They generally stayed on task, working diligently to reach their joint negotiated goal. They strayed from academic language as they paused to ponder the fat content of baking powder and butter (lines 22–27), and later, when Jennifer mentioned her hair (line 30). Although these side conversations were off topic, they helped to solidify the mutual engagement necessary for the bilingual girls to learn math.

Nestor

Unlike Jennifer and her girlfriends, Nestor was usually alone in math class. Although Nestor claimed that his favorite class in school was math, his below average performance did not reflect his enthusiasm for the discipline. His grades were generally low and he often failed his in-class tests. He didn't talk loudly, nor did he talk much in math class, but he still felt there was a possibility he could have been a top student in class. When I asked him if he thought he was one of the best students in the class, he said maybe. Upon hearing this, Jennifer quickly rebutted: "You barely even talk!"

Although he was from the U.S., Nestor had attended summer school in Mexico several times and noticed a palpable difference between math in each country. He felt that there was an emphasis on the individual in Mexico, as well as an emphasis on discrete skills (over discussion). Since in Mexico, math was done individually, he did not have much experience working in groups to solve math problems. He later informed me, though, that he believed his experience with math in Mexico helped him to understand it better here in the U.S.

Nestor sat up front near Ms. Loesely. He often seemed sleepy in class, as he moved rather slowly, and had a habit of resting his head on his desk and slouching. Ms. Loesely often reprimanded him, as she did others, for the 'Club Med slouch'. In addition, she often appeared frustrated with Nestor's sleepiness and inattention, as illustrated on January 29 when she asked him a question that he didn't respond to. To get his attention, she yelled: "Nestor to earth!" He was also often absent from class. His quiet and sleepy demeanor, coupled with his frequent absences, caused Nestor to be overlooked by others. He did not have many friends in class, except for a tall, good-natured, English-speaking blonde kid named Scott who was mainstreamed from special education classes. Nestor often worked with Scott in partner and small group activities, but their relationship did not truly solidify until the latter part of the spring semester. Because they sat at the front of the class next to each other, Ms. Loesely often helped them with their math work. Besides Scott and Ms. Loesely, Nestor wasn't often spoken of by others in class.

The majority of Nestor's interactions in class were characterized by quiet responses and self-talk. I could hear him working out problems orally on the days he was wearing a microphone. He often participated in whole class discussions, but usually under his breath. Later in the semester, it seemed Nestor's comfort level raised and he began to participate more often, both in whole class discussions, and when working with a partner, such as Scott. In Example 3.9 from late in the spring semester, the class is reviewing conversions of fractions and decimals. Nestor follows along with the teacher and is an active participant in the lesson.

EXAMPLE 3.9
NESTOR WAKES UP? MAY 10

L = Ms. Loesely
N = Nestor
B = Boy

1. L: The numerator goes in the division house, the denominator goes out, this should be review!
2. L: Now we've got a decimal, what do we do?
3. L: Move the decimal <two places to the right> [>].
4. N: <two places to the right>[<]!
   (in a strong, loud, confident voice)

———

5. L: Okay, now, two-fifths.
6. L: What goes in?
7. N: Two!
8. L: What goes out?
9. N: Five!

———

10. L: What does percent mean?
11. B: I dunno, total?

———

(Teacher asks students to look up the meaning of percent in their textbooks)

12. L: Percent means how many out of a hundred!
13. L: So anytime you have the whole thing you have percent.
14. N: Cool!

———

15. L: How do you turn a percent to a decimal?
16. L: you move the decimal <two places to the left> [>]!
17. N: <two places to the left> [<]!
18. L: If I turn point thirty-seven to a percent, I move my decimal...
19. N: To the right!

Students in Ms. Loesely's class were accustomed to her common quips such as 'the numerator goes in the division house' and 'move (or march) the decimal point over two places to the right'. These oral algorithmic routines, which she repeated often, were an important part of the shared repertoire in her class. On May 10, Nestor demonstrated that he had been soaking in Ms. Loesely's teachings over the semester. His unprompted responses in lines 4, 7, 9, 17, and 19 show that he had been listening when his teacher taught them how to convert decimals to percents, and percents to decimals.

On this day, the class was just reviewing concepts they had learned back in January and February, but they, along with the unusually enthusiastic Nestor were engaged in the group negotiated goal. Nestor's interest is obvious in line 14 when he discovers that he can use percent in a context in which he may not have yet thought of. Upon learning the definition of percent, he exclaims "Cool!"

At this late point in the semester, Nestor's identity has shifted from a sleepy and reserved student of math, to a more engaged participant. He was, without a doubt, moving toward full membership in the classroom CoP. Success in school, from the perspective of teachers and administrators is, however, primarily measured by achievement on formal assessments. Unfortunately for Nestor, time was not on his side. Much of the semester's grading was complete, and the standardized exam for the year was already complete. It was obvious that Nestor's quiet and sleepy exterior hid a learner who had much to absorb.

Much like Nestor was in Ms. Loesely's class, Anna was somewhat reserved in her math class at Ritter Middle School. Unlike Nestor, Anna was successful in math. In the last classroom, we will learn how focal students Anna and her clever and gregarious friend Pedro interacted in the community of practice which was created in Ms. Koch's second period math class.

Ms. Koch's Class

Of the three teachers in the present study, Ms. Koch was the strictest, most organized, and most loyal to CMP in its prescribed form. It was obvious that students in Ms. Koch's class learned early on to adhere to a very specific set of disciplinary criteria which included being in one's seat by the time the bell rang, sharpening pencils and/or preparing or locating any other required school supplies by the start of class, not getting up out of one's seat without explicit permission, and not talking unless the teacher had given specific instructions to do so. In some way, the structure and order of Ms. Koch's class struck me as counterproductive in a class which so faithfully employed CMP, a curriculum emphasizing discussion, interaction, and group work. However, Ms. Koch maintained that order and discipline were necessary for those classroom tasks to be effectively carried out.

Although she was strict, and some students indicated that they feared her, Ms. Koch was in general well liked by her students. In addition, she consistently produced successful students (in terms of classroom grades and standardized test scores) and was therefore highly recognized by the administration of Ritter and of the Springvale district. While she did appear to have favorite students (including Pedro) who consistently answered her questions correctly, Ms. Koch made a concerted effort to recognize all students in the class. In fact, she often asked her enthusiastic students to allow others a chance to participate. Thus, she tried to give equal time to each and every student.

Her very large and spacious classroom was arranged in six groups of four to five, so that all students faced the front. In total there were 22 students in her second period class. Since she did not speak Spanish at all, Ms. Koch grouped her bilingual students in such a way that the more proficient English speakers supported the limited learners in class. Ms. Koch worked to promote an atmosphere of shared cooperation and group work by arranging the students in small groups, each seated at their own table. In addition, these groups were generally quite stable. They were not often moved away from each other unless there was a necessity to do so. Ms. Koch also provided a very predictable classroom routine. Class was always begun with a short warm-up which was copied from the overhead projector or passed out; students were usually allowed to work together on it, and after it was completed and discussed in a whole class format, students were required to pass in the warm-up to a designated person at the table, who was responsible for getting them to the teacher. Next a large class discussion ensued, serving to introduce the class participants to the main lesson, and then Ms. Koch would pass out the required realia or manipulatives, and the necessary paperwork to carry out the lesson. At that point, the small groups would have time to do their work. Finally, the teacher would close the lesson with a wrap-up of sorts, drawing conclusions and tying loose ends together. The effect of this generally very predictable daily classroom routine on the students was that they knew what to expect from their class, teacher, and each other, and that they could be active agents in the participation of their class. Not only that, they were provided with a good deal of time daily in which to interact with each other.

Like Ms. Loesely, Ms. Koch had high expectations for her students and incorporated a regular routine with regard to the lessons and activities, rules and regulations, and development of important discursive mathematical notions. Ms. Koch had the daily lesson written on the board with the date, as well as the new and review concepts the students would encounter on that particular day. Ms. Koch used repetition and encouraged choral responses from the whole class to promote learning. For example, when the students were learning about percent, Ms. Koch would query almost daily: "What does percent mean?" To which the students would respond chorally "Out of one hundred!" When she elicited the meaning of probability, students knew to respond: "chance!" In terms of classroom order, she regularly reminded the class to raise their hands, to respond to questions thoughtfully, rather than impetuously, and to obey all classroom rules, including respecting each other. Ms. Koch was precise and consistent in ensuring routine in her classroom. She took full advantage of learning opportunities in class.

Ms. Koch's Class as a CoP, February 13

On February 13, Pedro and Anna were seated at their regular table with tablemates Emily and Sarina. The day's lesson revolved around games of chance as related to probability, so the teacher gave each of the four group members a cup with a certain number of colored cubes and the individual group members were instructed to do 25 trials where they selected 2 blocks to find matches. The match and no match results were to be recorded on their group tally sheet, and the final individual total was to be written as a fraction over 25. When added together, the total fraction of Anna and Pedro's group had a denominator of 100. In the following example, the teacher has elicited the data that the small groups came up with, and is now encouraging students to estimate the whole class's data which found a match, then asks them how to determine the whole class's data set for 'no match'. At the beginning of Example 3.10, Ms. Koch has just asked the students how many times 525 can go into 3250, as she was adding up several tables (and added a zero), but then she decides to round the numbers. When a student yells out "two," a clearly unacceptable choice, Ms. Koch corrects her and redirects the discussion.

EXAMPLE 3.10
KOCH: EXPERIMENTAL PROBABILITY AND FRACTIONS, FEBRUARY 13

K = Ms. Koch
G = Girl
A = Anna
P = Pedro
B = Boy

1. K: Well let's wait before we say something, <let's just take a minute> [>], five hundred times two would be one thousand, and I'm wanting to know how many times it would go into three thousand if I round it to three thousand, Donna, we think about six times, is that what (Donna doesn't answer)
2. A: <Six, six> [<] I know! (self-talk)
3. K: Let's try six and see what happens, what's six times five?

4. B: Thirty.
5. P: Thirty.
6. A: Thirty. (self-talk)
7. K: Six times two?
8. B: Twelve?
9. K: Plus three?
10. B: Uh, fifteen.
11. K: Fifteen.
12. K: Six times five?
13. G: Thirty.
14. K: Plus one?
15. P: Thirty-one.
16. K: Okay, so we've got pretty close that time, didn't we?
17. K: We're gonna stop right there...my no matches have to be done.
18. K: Let's see if you can figure that out without changing it to a decimal.
19. P: The denominer, the denominator is still one hundred twenty-five, right?
20. K: Yes, but I don't want to change this to a decimal, I just want to look at all the information I can and see if I can figure out an easier way.
21. K: What do I know about probability?
22. G: It's chance.
23. K: I know it's chance, but... what does it have to be when I do an experiment, if it's impossible, it is?
24. A: Zero. (quiet, self-talk)
25. K: If it's a sure thing it is?
26. A: one. (quiet, self-talk)
27. K: And the number one as a decimal is one decimal point zero zero, what is it at the percent?
28. A: <Hundred percent> [>]. (self-talk)
29. P: <Hundred percent> [<].
30. K: So if I know what this says as a percent, could I very easily figure out what this is as a percent, if I know that the total is one hundred percent?
31. K: No, I'm not gonna do it in my head, I'm gonna use my paper and I'm gonna figure it out, even I, I don't do this stuff in my head.
33. K: Okay, I have five extra points for whoever can tell me who gets it...

In Example 3.10, Ms. Koch has involved a number of classroom participants in the discussion of estimation and probability. She employs the traditional IRE (initiate, respond, evaluate) pattern in her whole class discussion, but does so strategically. She initiates (I), students respond (R), and she then evaluates their input. Line 1 reveals how Ms. Koch's redirects Donna, one of the students, as well as the whole class to focus on her question and to truly give thought to the answer they will provide. Although only some of the students are participating aloud in this group discussion, many, including Anna, are following along quietly, answering questions as self-talk. Anna carries out this self-talk, quietly under her breath in lines 2, 6, 24, 26 and 28. The complete utterance reflected in line 1 shows Ms. Koch interacting with Donna in particular, but appears to ask a general question on the part of the whole class, using the plural 'we' rather than 'I'. Anna indicates that she knows the answer that the teacher elicits, but does not get recognized (line 2). Nevertheless, Anna's input indicates that she comprehends and is engaged in the class discussion and is focused on the group task at hand. Lines 3–16 exemplify a rather common occurrence in Ms. Koch's class. Often, as in the above example, Ms. Koch would draw in the students by eliciting oral calculations. There were usually several students who would answer the mathematical problems she posed aloud, but more commonly, the students would mouth the answers to the questions from their seats. Here, four students, including Anna and Pedro, are participating orally in the discussion, but in actuality, many more are drawn into the interaction and are engaged in the group negotiated discussion.

Lines 16 and 17 show Ms. Koch shifting the discourse from a joint, oral calculation to a new topic. She explains that the group's focus will now shift to the 'no match' data from the group experiments, but that they will not change the fraction to a decimal (lines 17, 18, 20). Pedro is clearly involved in the teacher's focus. By clarifying her directions (line 19) he makes it known that he is still lacking direction, but is nonetheless following her lead. He is also taking the opportunity to practice a very important mathematical vocabulary word, denominator. Although he trips up on it at first, Pedro gains command of the pronunciation the second time. Ms. Koch takes a cue from Pedro's questioning and decides to refocus the discourse to make sure all are attending to what they need to be attending (lines 20–27).

Ms. Koch's use of IRE sequences as a strategy is seen in the following lines. She directs the students' mutual attentions to a joint negotiated enterprise (probability) by repeating concepts from third period's shared repertoire of mathematical discourse (lines 21–30). Ms. Koch elicits the participants' knowledge about their common repertoire regarding the meanings and uses of the terms probability, chance, possibility or impossibility, and 'a sure thing' (21, 23, 25, 27). Although she happens to be speaking very quietly and seemingly to herself in lines 24, 26, and 28, Anna answers the teacher's questions promptly, indicating she is highly engaged in the classroom discourse. Pedro also responds with Anna to Ms. Koch in line 29, indicating his involvement in and comprehension of the task.

After the initiation (I) and response (R) occurred, Ms. Koch carried out the evaluation (E) phase of the lesson. She evaluated students' responses by either continuing her questioning (as seen in lines 3–16), thereby implicitly crediting the responses as positive; revoicing their input (lines 10, 11); or by her consistent use of the 'I' referent, indicating what she would do (lines 20, 21, 23).

Ms. Koch shifts the IRE pattern she started with by wrapping up the sequence with a reminder to the students of how they should be carrying out the calculations (line 31). They should do calculations on paper. Even Ms. Koch maintains that she—a teacher of mathematics—does not do math problems like the one they have now in her head, promoting the message that teachers are also learners and have to work to solve problems as the students do. Her identity as an 'expert' math teacher is not static, but active. She sees herself as also a novice learner and experimenter, and clearly wants her students to see themselves in this way. Ms. Koch closes the discussion by providing an incentive of five extra points—likely to encourage engagement in the joint enterprise she has presented to the group.

Later in the class, we learn that Sarina has gotten the answer correct and is poised to earn the points, but Anna protests, out of earshot of the teacher, that Sarina has copied her. This interesting example, which we will return to in the next section, illuminates how Anna's identity plays out in math class.

Anna

When I asked her how she felt about math, Anna said she liked it, but reading was actually her favorite class in school. She explained her feelings about math: "sometimes I don't like math cuz it's boring, but sometimes I do like, no just when we play games, when I learn new things." Two major obstacles for Anna in math class were her comprehension and production (of language). On one hand, she couldn't understand her teacher, and on the other, she feared speaking aloud in class. She explained her reticence:

> I, I, don't even raise my hand cuz, when I was in the other semester, it was some girls and boys that laughed on me when I speak English, and it was bad,

> and that's why I don't raise my hand anymore, I, I, know how to raise it, but I, I don't allow, cuz then they laugh of me.

Her quiet character and shy demeanor caused others to overlook most of her contributions in class. She was hesitant to speak up in class not only because she feared ridicule from her fellow students, but also because she was not fond of her teacher. She was afraid of Ms. Koch and felt she was mean. Ms. Koch did call on her occasionally in class, and she was an active participant in small group activities, however, Anna was not recognized by her fellow students as one of the very best in class. Some of her usual group partners, Pedro, Emily, and Kathryn, admitted that Anna was good at math, and named her as one of the people who they worked well with on a daily basis.

Despite her timidness, Anna was a successful student. Above average grades, preparedness, punctuality, and dependability contributed to her 'good student' identity in school. Through quiet self-talk recordings in math class, I became aware that Anna was capable of making mathematical calculations in her head and on paper.

Her identity as a 'good student' was also seen in interactions with her group mates. She often played the role of a mother hen, admonishing her friends to get on task and stop playing around. On February 10, the activity of the day was to roll sets of numbered cubes to determine whether a match was made between the two. On this day, Anna commanded her partners to "Go!" "Wait!" and "(Emily), stop playing around!" in a forceful tone.

She felt quite comfortable bossing around her friend Pedro. On February 23, Anna redirected him with a stern "Get busy, Pedro!" On February 12, when Anna noticed Pedro was calling out answers without adhering to classroom decorum, she snapped "Raise your hand!" But just as easily as she would reprimand Pedro, she could be heard eliciting his assistance, as in her plea "Pedro, help me." The shared relationship between Anna and Pedro was integral to Anna's math learning. Since he was one of the few people with whom she could speak Spanish, her strongest language, she valued the partnership and support that he offered her when he attended Ms. Koch's second period.

Despite limited English proficiency, Anna was able to participate in the negotiated enterprises in which her class was involved. She understood well the shared repertoire of knowledge and resources that characterized Ms. Koch's second period. But her hesitation to engage in whole class discussions occasionally triggered negative consequences, as in Example 3.11 from the end of the class period on February 13. Here, Anna is engaged, but participates quietly.

EXAMPLE 3.11
YOU COPIED ME! FEBRUARY 13

K = Koch
A = Anna
S = Sarina
J = Jean

1. K: Okay I know that the probability # if it's impossible, it's zero, I know that I'm going to get two cubes out of that cup, because that's all that's in there, all I picked was two cubes, that was a sure thing, probability is the number one, one as a percent is one hundred percent, now if I know that there is nineteen and six-tenths percent, [?] I know the total has to be one hundred percent.
2. K: Jean, what would it be?
3. J: um.
4. A: Eighty and four-tenths. (whispering)
5. K: Charlie, what's it going to be?
6. S: Eighty and four-tenths!

———

7. A: You copied my answer, Sarina!
8. A: See ah, see my answer, right?
9. A: I know # probability.
10. S: I had it but I didn't think it was right.
11. A: Miss, I know, knew the answer, but I don't like to speak up.
12. K: Well, why don't you like to speak up, if you know the answer?
13. A: Look, cuz I get nervous, Miss, I answer, I can't do it sometimes if I, If I'm like ah!
14. K: [=! laughs] If you have the right answer, don't sit there and say you're afraid to answer.
15. A: No cuz, one, time I was in there and they laugh of me when I talk.
16. K: Well, that's alright.
17. K: Look, you know two languages and that means you're a whole lot smarter than a lot of people that just know one.

This example shows how Anna's timidness in class cost her extra credit points and potential recognition by the teacher. After Ms. Koch has presented the problem for her students to solve (line 2), she calls on Jean, an overachiever who could normally be counted on to provide the correct solution. During this time, Anna reaches the correct solution and whispers it to herself (line 4). As Ms. Koch passes the question to Charlie, Sarina takes the opportunity to get credit for the solution and yells it out to the class and teacher. Sarina's confident outburst (line 6) is recognized by Ms. Koch, and she is then honored with praise and the extra 5 point prize.

A few minutes later, Anna drums up the courage to confront Sarina with an accusation that she had copied her (lines 7, 8). Sarina counters that she actually knew the answer but lacked the confidence to reveal it to the public (line 10). Anna then engages the teacher to set the record straight—that she was responsible for the correct answer for which Sarina was recognized (line 11). However, Ms. Koch did not offer the comfort Anna was seeking. Instead Ms. Koch questioned Anna's reluctance to speak, and Anna provided her rationale as to why she wouldn't speak (lines 13, 15). At the end of the interaction, Ms. Koch absolves Anna for her reluctance to speak up and credits Anna for her bilingualism (lines 16, 17).

Although conflict arises out of the class discussion, it is nonetheless evident that several participants, namely Anna and Sarina, have been mutually engaged in the academic enterprise that was the focus of the lesson. Anna's identity undoubtedly plays an important role in her participation in the class practice. As well, Anna's engagement with her teacher and her classmates contributes to her math learning and interaction in class. These girls' interaction reveals how they, along with other class participants share in common a mutual engagement, a common repertoire, and focus on a joint negotiated enterprise.

Pedro

Math was Pedro's favorite subject because according to him "It's easier to learn." He didn't think the concepts he was learning in math class were hard and he got relatively good grades without much effort. When I asked Pedro if he thought he learned more in groups when he could receive help from others, he indicated he knew more than most others in the class. "I already know the more" he said. And on February 13, when doing the match experiment, Pedro refused to accept advice or suggestions from the bilingual assistant in the classroom, Mrs. Hammer. He knew she wasn't good at math (which Ms. Koch later confirmed to me) and would speak only to Ms. Koch to get his question answered. He normally worked in the same group with Anna, as they sat at the same table. In both whole class discussions and small group work, he took a leadership role. He was enthusiastic and liked to command the teacher's and students' attention. He was a go-getter, to be sure.

Although Pedro only attended Ms. Koch's second period math class intermittently, he was an integral player in the CoP which emerged in the class. Pedro was quite active and participatory in whole class group discussions.

Because he had been born in the U.S. and educated in U.S. schools, Pedro was comfortable using English. His experience and knowledge helped him in his role as a helper and friend to others, but especially Anna. Although he was still considered a novice in terms of language learning (according to school designations), he sometimes shifted between playing the role of expert and novice, in particular when he worked with Anna.

Since they were in ESL together, Anna and Pedro had established a strong, supportive relationship. Although Anna was Mexican and Pedro was (Mexican-) American, they both valued the American system of education and recognized the benefits that academic success can bring. School was very important, and while they both had friends and were considered (somewhat) 'cool' (according to some of their friends in class), they felt it was important to maintain and nurture their academic identity. For example, Pedro was not ashamed or embarrassed when the teacher would single him out for good work in class, on test preparation activities, or classroom tests. Because she was shy (and because others made fun of her), Anna would become a bit embarrassed. However, although she was just a B/C student, she made it clear to me that she was very academically oriented and that school was highly important to her.

Not only was Pedro bright, gregarious, and a good friend to Anna, he was also very enthusiastic about math. He also had a good rapport with Ms. Koch. She often picked him to answer questions because he was consistent in producing the answers she was looking for. Example 3.12 occurs on one of Pedro's early visits to Ms. Koch's class. Today, February 12, the class is studying probability from the CMP textbook entitled *How Likely is it?* and are discussing whether a game called 'Evens and Odds' is a fair game. She is asking if there are prime numbers on each of the rows of the game board.

EXAMPLE 3.12
PEDRO'S ONE MAN SHOW, FEBRUARY 12

K = Koch
P = Pedro
G = Girl
A = Anna

1. K: Okay are there any prime numbers on the second row? (someone says no)
2. K: Are you sure?
3. P: Two.
4. K: Two.

5. K: By the way, two is the only even prime number xx, okay let's look at row three, are there any prime numbers there?
6. P: Three.
7. K: Three . . . so are there any other primes in row three?
8. P: No.
9. K: How about, are there any other primes in row four?
10. P: No.
11. K: All these are multiples of two aren't they?
12. P: Uhhuh.
13. K: Okay, row five.
14. P: Five.
15. G: Five.
16. K: Five, and what about row six?
17. P: They're all +/.
18. K: They're all what kinds of numbers, if they're not prime, what are they?
19. P: Composite.
20. K: Very good, composite, you remembered that from a long time ago, that's good.

—

21. K: So then the probability of a prime, xx what do I get?
22. P: One sixth.
23. K: divide by what?
24. A: Six.
(quietly, but loud enough so the teacher could have heard)
25. K: And I get one +/.
26. P: Sixth.
27. K: Sixth, okay good.

Example 3.12 illustrates Pedro's engagement in the whole-class class discourse. Lines 3, 6, 8, 10, 12, 14, 17, 19, 22, and 26 show how Pedro not only participates in the discussion and engages in one on one interactions with the teacher, but also how he is consistent in getting answers correct. Ms. Koch recognizes Pedro's contributions and credits him with acknowledgment and praise. Indeed, Pedro the go-getter succeeds in leading a one-man show.

More than once, Pedro's enthusiasm resulted in redirection by Ms. Koch. A representative example of this redirection occurred on February 26. On this day, Pedro finished a calculation before anyone else in the class and thrust his hand in the air. Because he had answered many of Ms. Koch's questions earlier in the lesson, she told him: "Pedro, let's let someone else answer, I am very glad you are answering, but you are very fast today." Upon being reproached, he usually quieted down for several moments, only to brighten up again shortly after.

When I asked Pedro whether he felt being a boy helped him in math class, he responded in the negative. But being the only male in a group of girls (throughout the semester) had an effect on him. He was usually quite respectful of his female partners, but he used gender to his advantage on a later date. On March 23, Pedro wasn't being recognized in the joint negotiated goal he and his group mates were working on. The task they were assigned was first to order a meal, drink and dessert, if desired, from a menu supplied by the teacher. Then they would total their order for the table and calculate the tax and tip percentages, and then separate the bill by guest. His partners, Anna, Emily, and Sarina were discussing who would take the orders and who would give the orders first. Frustrated that he was left out of the decision making, he barks: "Hey, no, you gotta get the, me, cuz I'm the guy in here, I gotta get the expensive one!" Upon hearing this, Emily changes the subject by telling Pedro that his meal selections are disgusting. Thus, playing the gender card did not help Pedro succeed in accomplishing his goal of being recognized.

Pedro's identity as a good (and favored) student, helper, ELL, and all around go-getter facilitated his success in math class. His role as a visitor in Ms. Koch's second period did not detract from his ability to comprehend and engage in the shared repertoire held by the CoP in class. He was an engaged participant in teacher-mandated academic tasks and even served to support the learning of his more limited English proficient colleague, Anna. While Pedro still had a good deal to learn with respect to mathematics and English as a Second Language, he was an active and successful participant in the CoP of Ms. Koch's second period at Ritter Middle School.

CONCLUSION

In the examples of classroom discourse presented above, I drew from both teacher-fronted and small group student discussions to show how each of the three math classes I examined worked as a classroom CoP. The goal of each of the classes was to promote the development of mathematical and academic knowledge, and while all of the three classes were successful in achieving this goal to some degree, some of the classes were more so. I have revealed how the three classes, while different, all exhibited characteristics of a CoP. All three classes shared a mutual engagement, and a shared repertoire, and all were working on some type of joint negotiated enterprise—in the case of the examples—dealing with fractions and probability in CMP lessons.

The identities of each of the six focal students I followed were diverse. Thomas was considered to be a math whiz; Benjamin was unsure of himself, responding to teacher queries with 'I dunno'; Jennifer was responsible and mature; Nestor was a sleepyhead; Anna was considered shy and timid; and Pedro was a go-getter in math class. Each student was an important participant in their communities of practice, but their individual identities affected their learning and ultimate success in the classroom. Student identity coupled with teacher identity and features of the joint practice that emerged in their classroom communities also contributed to student success.

Each math class CoP had its own impressive and effective features. Mr. Martinez' students seemed to exhibit a feeling of equality with their teacher. Since he spoke Spanish and joked around with his students, participants were drawn into the lessons. In Mr. Martinez' class, Thomas played the role of the expert as teacher and apprentice, in addition to the role of student as novice, common in K–12 education. Benjamin was able to rely on his knowledgeable friend, Thomas, who could not only help him understand the math, but also the language if necessary.

In Ms. Loesely's class, there were also strong bonds between teacher and students. The personal connections shared in the class, like that which emerged from discussions of Bill, the class fish, served to strengthen the shared (social) repertoire of the group. The diversity of the format of Ms. Loesely's class, as well as the group activities also enhanced the engagement of the class. Ms. Loesely's use of choral responses helped solidify concepts, and definitions her students would apply in class. Jennifer benefitted from group work when she was able to interact with her bilingual classmates. Her maturity and thoughtfulness was a good fit with the open-ended discussions that were encouraged within the CMP lessons Ms. Loesely introduced. Nestor was well suited to whole class discussions as well as individual interactions with the teacher.

Finally, Ms. Koch's classroom was very predictable. Her students understood the daily routines and expectations in math class, and as a result, knew what to expect from their teacher, their class, and each other. Ms. Koch did not speak Spanish, but for those limited learners of English, like Anna and Pedro, the consistency of Ms. Koch's class assisted them in comprehension of general academic and mathematical concepts. Her seating arrangement and strategic grouping allowed for interaction between bilinguals. Further, Ms. Koch's emphasis on repetitive, choral learning of specific lexical items and short definitions also appeared to be useful for not just the ELLs, but for many of the sixth graders.

Some features of each class were, however, not conducive to learning for some ELLs. Mr. Martinez' class was noisy, busy, and wholly teacher-centered. Students such as Benjamin, who needed focus and individual attention, did not always get it. Ms. Loesely's class was active and diverse (in terms of structure and daily activities), and thus lacked the routine that many of her students needed. Ms. Koch's class provided stability, structure, and consistency, but in a strict, rigid, and rule-governed environment. Learners like Anna, and in some cases, Pedro, struggled to interact with their teacher and other students in such an environment.

Each of the three classroom communities of practice engaged in practices. Some of these practices supported learning, while others appeared to have inhibited it. In addition, students' individual identities contributed to their success in class. But the question must be asked: What is the effect on learning on individual students when diverse identities mix with the situated practice in each classroom community? To begin to explore this question, we examine the successful and unsuccessful focal students in the three classes, and contrast them in terms of their classroom practice.

Anna and Benjamin were both relatively recent arrivals, having immigrated to the U.S. two years earlier. They were both quiet and shy and somewhat apologetic about their lack of English skill. Both were Spanish dominant, enrolled in ESL, and both received language support in some way in their math classes. Anna was relatively successful and Benjamin was not. Pedro and Nestor were both born in the U.S. They both had ties to Mexico and returned often. Both spoke Spanish at home, both had been in language classes throughout their academic career and in fact, were both still in ESL at sixth grade. In addition, both of these boys played the role as a helper to a less advanced learner in their classroom. Pedro was successful, but Nestor was not. Finally, Thomas and Jennifer were both born in the U.S. but were considered 'fluent English proficient' by the school system. They had both been exited from language support classes within the past few years and were mainstreamed into all English classes. They were both outgoing and highly engaged in math class. Thomas was extremely successful, but Jennifer was not terribly successful.

Why did Anna, Pedro and Thomas succeed on school-based measures of success, while Benjamin, Jennifer, and Nestor did not? Certainly factors such as their prior experience in math, quality of (prior) math and language courses, motivation, and possibly aptitude played a part, but their relationship to the practice which emerged in their math communities did as well. Although Anna by no means thrived in Ms. Koch's strict environment, she enjoyed the closeness and interaction she shared with Pedro and her other bilingual partners, who could give her the assistance she sometimes needed. Benjamin could ask Thomas or the teacher for help (in Spanish or English), but because the format of the class was usually teacher-fronted, they were few opportunities for him to do this. Pedro was a good fit for Ms. Koch's class.

He was able to demonstrate his knowledge in large group settings, but could participate in small groups well. Jennifer benefitted from group work in Ms. Loesely's class, and was quite suited to the open-ended questions that CMP provided. But Jennifer's group and individual work was often not monitored. She needed more structure and more guidance by the teacher as well as from more capable peers. Thomas was well suited for Mr. Martinez' class. Group work wasn't the most effective learning environment for him because he was independent, self-motivated, and fast. He enjoyed being the class whiz kid. The last focal student, Nestor, seemed to fall through the cracks. He needed more time to absorb content and learn the concepts. Ms. Loesely's use of repetition helped him to remember routine math concepts, but he would have likely profited from much more strategy learning.

From this comparison of three very different classroom communities, and six quite different focal students within them, it is evident that while each may contrast strongly from another, each functions effectively and appropriately as an independent community of practice. Language learners in each of the three communities learned math and math discourse, however, their membership as a peripheral or full participant was dependent on their participation and engagement in class.

Roles of novice and expert were dynamic and played out in varied ways within classroom participation. In these communities, some focal students exhibited a great deal of agency (Thomas and Pedro). These focal students interacted with others as teachers and experts, in addition to playing the novice learner in class. Sometimes, even teachers played the role of learner (Ms. Koch and Mr. Martinez), as when Mr. Martinez told students he did not know if the answer to a problem was in lowest terms (Example 3.5), or when Ms. Koch expressed that even she couldn't do math problems in her head (Example 3.10). These examples of teachers as novices problematize the traditional conceptualization of the novice/expert dichotomy.

The next chapter will further explore how focal participants interacted as a learning community and were socialized to develop mathematical discourse in reform-inspired classes. In chapter 4, I will show how participation frameworks developed within fraction lessons framed as play and will also illustrate how frames of play within CMP served to engage and motivate some ELLs in fraction lessons.

NOTES

1. Ms. Loesely as well as Ms. Koch mentioned mathematics curricula entitled AIMS and GEMS when referring to CMP influences.
2. Fraction kits were a lesson from the CMP book, *Bits and Pieces I* which consisted of several strips of colored coded paper that indicated sizes of fractions, i.e., one-half, two-thirds.

CHAPTER 4

FRAMES AS PLAY AND PARTICIPATION FRAMEWORKS IN REFORM MATH

For many years, the notion of framing talk or discourse has been incorporated in linguistic (Schiffrin, 1994; Tannen, 1993), psychological (Rumelhart, 1975), sociological (Goffman, 1983; Hymes, 1974), and even cutting-edge political literature (Lakoff, 2004). In this chapter, I employ the different, but nonetheless related,1 concepts of frames as play as associated with education. As used in this chapter, frames are interesting and motivating, but imaginary, or 'play' scenarios used to help learners understand and engage discursively in math. I have termed them frames of play as a result of their connection to the imaginary.

In this chapter I reveal how the six ELLs I followed used frames of play to learn math, and show how the students worked together within participation frameworks to discuss math-related topics. Participation frameworks are the various roles, responsibilities, and constellations of alignment (O'Connor & Michaels, 1996) taken up within mutual engagements between negotiating participants. While neither the integration of frames into language education literature (see Lytra, 2004, 2008), nor specifically into math education literature is an innovative concept (see, e.g., O'Connor &

English Language Learners and Math, pages 67–94

Michaels, 1996), here, I will reveal insights about how framing is used in reform-oriented mathematics classrooms to achieve a number of beneficial learning outcomes for both bilinguals and second language learners alike. In particular, I explore how frames contextualize mathematical content for ELLs and work to encourage participation within the mathematical communities of practice in each of the three classes.

The participation frameworks that emerge within classroom activities tie into the CoP. An examination of participant frameworks reveals the various roles that learners embody when discussing math, and elucidates how shifting identities contribute to, or hinder socialization to mathematical discourse and math learning. Examining types of participation frameworks can provide evidence as to opportunities for engagement in class. In addition, the CoP model holds that participation in shared practice is equated with learning. Participation frameworks can help explain how learners do or do not become part of mutual engagement within a classroom setting, and can thus reveal something about how learners do or do not learn math.

FRAMING ACADEMIC DISCOURSE

In this chapter, I employ the concept of framing to gain an understanding of how the discourse of play facilitates the development and production of academic mathematical discourse. Within math class, frames of play are interesting, motivating, and relevant scenarios (such as game playing, brownie making, or money handling) that, in many cases, contextualize math while promoting reasoning, problem solving, arguing, and rich, engaged, focused academic discussion in class.

Frames of play provide ELLs in content courses like mathematics a space for transforming reality (Wohlwend, 2007) (or for transforming the traditional math class) to an imagined or play setting, where they can interact in and about math in an interactive, playful atmosphere. Frames of play thus provide ELLs the forum within which to engage in and about math.

FRAMES OF PLAY

Frames are considered the mechanisms that speakers utilize to structure their interactions, experiences, and responses to events (Goffman, 1974, pp. 10–11). Bateson (1972), who coined the term 'framing' (Tannen & Wallat, 1993) as it is used in social science, maintains that participants can differentiate between playing and other types of interactions (i.e., fighting), provided they understand the metacommunicative cues used within the frame. Building on Goffman and Bateson's definitions, Wohlwend (2007) describes frames of play "as a space for contextualizing metalinguistic propositions or for exploring hypothetical premises by transforming assertions into concrete, albeit imagined contexts" (p. 6). Frames of play can then be understood as both an impetus (frame as context or scenario) and a result (the ensuing discourse) of a linguistic interaction.

Vygotsky (1978) recognized the importance of play. Believing that the social environment was critical to learning and development, Vygotsky (1978) stressed that during play, children and adolescents must maintain their roles within the particular play scenarios in which they interact. For Vygotsky (1978) play provided the foundation for the development of skills critical to children's adult life: in the intellectual, moral, and activity realms. Play has a purpose, according to Vygotsky (1978), and play ultimately leads to development. The frames of play discussed in this chapter are thus conceptually derived from Vygotsky's original work.

Hoyle (1993) describes framing (using Garvey's (1977) terminology) as a 'nonliteral orientation'. In her work, Hoyle examines the sophisticated changes in both the frames and the identities through which young, preadolescent boys shift when sportscasting. My use of the concept of framing as a nonliteral orientation is informed by Hoyle (1993). As used in this chapter, frames of play are the interesting and motivating, imaginary, or 'play' scenarios used to help learners understand, as well as the ensuing discursive interactions in math which surround the scenarios. I have termed them frames of play as a result of their connection to the imaginary.

In my research, as in Hoyle's, frames serve as a thematic scenario in which participants interact. Hoyle examined only one: the frame of sportscasting as play. In math class, a variety of frames are utilized. The frames serve as a nonliteral orientation that provoke discussion, interaction, interest, motivation, and problem solving in math. Frames are used in math as both a teacher-constructed device, as well as a convention of the CMP curriculum, to contextualize the lessons and promote academic discourse use.

"A 'presupposition' (or assumption, or implication, or background expectation) can be defined very broadly as 'a state of affairs we take for granted in pursuing a course of action'" (Goffman, 1983, p. 1). This definition from one of Goffman's seminal works on framing in social interaction posits that we as social beings presuppose that our interlocutors will hear and understand what we have to say, in the way that we mean. When a presupposition is implicit in a learning situation, learners must accept or 'buy in' to the presuppositions inherent within the academic task they are given, if they are to fully understand the task. For frames of play to be successful with ELLs, the learners must also buy into the frames of play presented to them in math.

LANGUAGE SOCIALIZATION

Part of "buying into" frames of play is being socialized to and through language. When ELLs listen to, interact with, and learn from teachers and peers (not that these actions are mutually exclusive), they are being socialized (but of course, also socializing others) to language of school. In math class, learners are being socialized to math discourse—the language of math they need to know to be successful in the classroom.

Language socialization was first used in linguistic anthropology as a way of explaining the processes by which young children learn to use language appropriate to the norms and values of their home culture (Schieffelin & Ochs, 1986), however, it has been fruitfully utilized in recent years to understand how school-aged children develop a second language in educational settings (He, 2003; Willet, 1995). Language socialization highlights the importance of a longitudinal approach, an ethnographic orientation, and a cross-cultural perspective (Garrett & Baquedano-Lopez, 2002, p. 341). Similarly, routines play an important role in the process of socialization to language (Corsaro, 1988; Garrett & Baquedano-Lopez, 2002; Kanagy, 1999). Routines, as I have described before, are useful for understanding how "learners impart and construct knowledge in that they show how both linguistic and conceptual knowledge develop over time" (Hansen-Thomas, 2008, p. 385). Thus, routines facilitate socialization.

In this chapter, I will present a variety of frames of play utilized in the three sixth-grade math classes that are incorporated into lessons as routine activities. Employing an ethnographic orientation and a cross-cultural perspective, I will both illustrate and describe how the six focal ELLs interacted in math class using frames of play.

FRAMES OF PLAY IN CMP MATH

Because CMP is designed to encourage interaction, problem solving, and communication, many of the group activities are framed within appealing scenarios. According to the guide to understanding the CMP curriculum, CMP authors Lappan et al. (1996) explain that "... the curriculum is organized around interesting problem settings—real situations, whimsical situations, or interesting mathematical situations" (p. v). Involvement in these frames is thus a critical component for students learning math within CMP.

Some of the play frames used in the sixth-grade math class centered around food, such as pizza and brownies; money, and sports, among others. These play frames are designed to function as a learning aid or tool—presented in the form of contextualization—to help students understand and engage in mathematics. These tools serve as scenarios to frame the mathematical concepts presented in the lesson. But not all scenarios used in class can be termed a play frame. In some cases, the scenarios included in the texts or in the teacher's discourse are just that: examples used to illustrate the mathematical concepts presented in the lesson and contextualize content. There is no doubt that contextualization is a key component of appropriate instruction for all learners, and for ELLs in particular. Research indicates that the use of authentic, meaningful activities can situate learning for ELLs and provide them the opportunity to make connections between what they know and what they are learning (Echevarria et al., 2008, p. 38).

So how does a frame of play differ from simple context or a scenario used to exemplify a math problem? I do not maintain that all word problems contextualized within a particular scenario (e.g., asking the probability of red to green gum balls, or using pizza to divide a fraction) is enough to be characterized as a frame. But when contextualization succeeds in sufficiently piquing the interest of learners to encourage them to participate actively when they would not otherwise, and produce academic language characterized by in-depth and higher order thinking, a scenario can be termed a frame. Specifically, the following criteria must be met: (1) students need to be engaged in the frame of play; that is, they need to buy into the frame, as demonstrated by their engaged discursive interaction and, (2) students need to be involved in the various roles that the frames of play provide. When these criteria are met, and students are participating as active, engaged students of math, I maintain that a frame of play is being utilized and created.

FRAMES OF PLAY IN CMP FRACTION UNITS

The units reflected in this chapter are those that illustrate fractions and fraction-related concepts such as percentages, ratios and multiples. The units which encompassed fractions in the sixth grade CMP textbook series[2] included *Bits and Pieces I* (Lappan et al., 2002a), *How Likely Is It?* (Lappan et al., 1998), and *Bits and Pieces II* (Lappan et al., 2002b). Within those three units, sixth grade learners were thrust into a variety of roles which framed the fraction-based lesson of the day. Those roles dictated by the curriculum included the following: builder, brownie maker/cook, gardener/planner, cat specialist, basketball/sports analyst, gambler/scratch-off card player, restaurant customer, and breakfast cereal analyst. Other roles that framed the sixth graders in their CMP math class included mathematician (one who measures, calculates, etc.) researcher (who conducts experiments and knows critical information like the outcome of an experiment in probability will be facilitated by an increased number of trials), team player (who works well in a group), and problem solver (who answers ACE (Applica-

TABLE 4.1 Frames in CMP Fraction Units

Mr. Martinez	Ms. Loesely	Ms. Koch
Bits and Pieces I	Bits and Pieces I	How Likely Is It? Bits and Pieces II
Sports/basketball analyst Mathematician	Mathematician Builder Brownie maker/cook Gardener/planner Cat specialist	Researcher Basket/football analyst Gambler/guesser of odds Restaurant customer Breakfast cereal analyst Mathematician

tion/Connection/Extension) questions in their book at the end of each unit section. Table 4.1 shows the frames used in CMP classes throughout the spring semester.

The frames used in the sixth grade CMP texts are used to encourage and motivate students to think about mathematical concepts in non-traditional ways. Thus, teachers following the CMP model were responsible for incorporating them into their classes. Speaking with Ms. Koch in an interview about CMP, I asked her what benefit she felt that the interesting problem settings, including games, themes, and frames brought to the learners. She said "It's more motivational... it's much more interesting," but maintained that she sometimes had to curtail the play that occurred within the frames: "I could really let them go wild... it's a fine line." Overall, though, she believed that the frames were a highly useful feature of CMP.

One strategy used in the CMP curriculum to assist in socializing students to these frames was use of the recurring theme. In *Bits and Pieces I,* themes of fundraising using fractions, making brownies or another similar dessert for a school contest or bake sale, dividing jelly beans, chocolate bars, or pizza to share amongst a class, discussing one's pet's characteristics (i.e., bad or good breath), managing money, or portioning land or space, such as a garden were regular, recurring themes. Designed to be used as a follow-up text to the preliminary fraction unit *Bits and Pieces I, Bits and Pieces II* reintroduced these same themes, but in different contexts.

The unit on probability, *How Likely Is It?,* reused and recycled many of the same themes, using the same characters and situations. In the first three chapters of *How Likely Is It?*, an eighth grade student named Kalvin is represented as the main character who initiates a whole host of activities using probability. Kalvin cleverly decides to employ probability and chance and makes a deal with his mother to determine whether he should eat 'Health nut flakes' cereal, or the more desirable, 'Cocoa Blast'. He later chooses a less random object to toss (marshmallows instead of a coin) to improve his chances of eating his favored cereal. By the end of unit three, the sixth graders have become quite familiar with Kalvin and his crafty schemes to use probability to his advantage.

In addition to the frames dictated by the curriculum, common frames of money, food, building, teaching/tutoring, and playing a variety of games were also often used in class. Like the CMP frames, these frames required astute learners to play a role related to the particular frame dictated by the teacher or lesson. Some of these other frames were introduced in conjunction with the lesson in the CMP text, but some were introduced through additional resources, or even through the teacher's experience.

TEACHERS' USE OF FRAMES IN FRACTION LESSONS

All of the teachers used frames of play from either CMP or from their own experience, however, some employed framing in their teaching more than others. In Table 4.2, I show the frequency and types of frames introduced by each teacher within lessons based on fractions and probability. Being a mathematician, researcher, problem solver, team member, or peer tutor, occurred often throughout lessons in both whole class and small group configurations. Sometimes these frames of play were directed toward individual students, rather than the majority of the class. Since these categories are more implicit in lessons than those in which a tantalizing topic such as pizza, candy, money, or basketball is introduced, I have only included here frames of play in which learners take on new roles within imaginary scenarios while learning fraction concepts.

Table 4.2 reveals that out of 9 days in which fractions were taught (and I was an observer in class), Mr. Martinez utilized frames of play 8 times. Out of 16 days where fractions comprised the primary lesson, Ms. Loesely used frames of play 15 times. Out of 27 class days in which fractions or probability were taught, Ms. Koch utilized frames of play 32 times. From this chart it is clear that frames of play were employed to assist students in understanding fractions. This chart does not, however, indicate the quality or manner in which the frames of play were employed, nor how students could use the mathematical concepts individually for standards-based evaluation.

When frames of play were employed in Ms. Koch and Ms. Loesely's classes, the frames usually encompassed the entire class lesson. This occurred because CMP grounds the majority of its lessons within a frame of play. When Mr. Martinez utilized frames of play, he often just referred to a routinely cited frame to clarify a confusing concept. Because Mr. Martinez did not use CMP much in his classes, frames of play were utilized to a lesser degree in his classes. Because Ms. Koch, and to a somewhat lesser degree, Ms. Loesely, used a good deal of CMP in their classes, these teachers incorporate more play frames in their lessons.

TABLE 4.2 Frequency and Types of Play Frames Used by Teachers

	January	February	March	April	May
Martinez (9 fraction days)			*	*	
Sports/basketball analyst	21, 23	9			
Food eater (pizza/candy)	28	4			
Engineer/Carpenter		4			13
Fundraiser	23				
Total references: 8					
Loesely (16 fraction days)			*	*	
Builder					
Food (pizza, brownie, candy) eater/cook	20, 22, 23, 28	2, 3, 4			
Gardener/planner		5			
Cat specialist		12, 16, 17			
Game player	21	11			
Money manager	22, 23				
Total references: 15					
Koch (27 fraction days)			*	*	
Basket/football analyst	16	17			5
Gambler/guesser of odds		13, 17, 26			
Restaurant/retail customer			23		3, 4, 5, 6, 7, 20, 21
Breakfast cereal analyst	14, 16, 20, 22, 27				
Game Player	29	3, 10, 12, 23			10
Money manager	16, 22	5			4, 7, 20
Cat specialist					10, 14
Total references: 32					

* March and April focused on standardized test preparation, so I attended less regularly

TEACHER-INSPIRED FRAMES

Because Ms. Koch and Ms. Loesely utilized the CMP curriculum most often, more of their fraction lessons incorporated frames than Mr. Martinez' did. However, during class, he, as well as Ms. Loesely and Ms. Koch, used a variety of what I call 'teacher-inspired' frames to help their students understand fractions. Pizza, fruit, money, and candy bars were common teacher-introduced frames used in all three of the classes.

Table 4.3 outlines additional frames used by the teachers.

TABLE 4.3 Other Teacher or Student Produced Frames in Fraction Units

Mr. Martinez	Ms. Loesely	Ms. Koch
Teacher/peer tutor	Game player	Game Player
Problem solver	Problem solver	Problem solver
Pizza eater	Pizza eater	Team member
Engineer/Carpenter	Team member	Money manager
	Money manager	

In the following section, I illustrate how some frames originating from the CMP curriculum and from teachers functioned as a motivating strategy to assist in the production of discourse by ELLs. I will first describe how Mr. Martinez, Ms. Loesely, and Ms. Koch routinely integrated frames of play in their classes over the course of a semester, and will then show how several of the focal students utilized frames of play in their respective math classes.

THOMAS AND BENJAMIN–FUNDRAISING, CARPENTRY, AND FRACTIONS

On January 23, Mr. Martinez' class has just completed an activity in which they created 'fraction strips'—various lengths of thin color-coded strips of paper that represent a variety of equivalent fractions (for example, one-half, two-fourths). They are going to use their strips to measure the fundraising success of the sixth, seventh, and eighth grade student body at Thurgood Marshall Middle school, a typical, but non-recurring make-believe example school described in Investigation 1 of *Bits and Pieces I.* The introduction for the chapter sets the stage by describing how each of the three grades was planning to raise money to buy band and sports equipment. Sixth graders were selling posters for their portion of the fundraising event. On page seven, an illustration depicts the sixth graders' progress on their fundraising goal from day 2 to day 10 in even numbers. Their progress is represented by thermometers which total $300.00, the sixth grade class's ultimate goal. The task accompanying this illustration requires students to use their strips to measure the Marshall sixth graders' progress based on the shaded part of the thermometer. Example 4.1 illustrates Thomas' engagement within the CMP lesson, as well as his shifting roles within participant frameworks created with the teacher.

In general, the students in Mr. Martinez' class were highly participatory in the classroom discussion that arose from the fundraising frame and its accompanying activity. Since the students were able to negotiate activity by using their own tools—the fraction strips—they were attentive and involved in the teacher-fronted discourse. As to be expected, Thomas (the good stu-

EXAMPLE 4.1
THOMAS: I KNOW HOW IT GOES! JANUARY 23

M = Mr. Martinez
T = Thomas
A = Armando

1. M: Okay, start trying them all (fraction strips), I want you to try em all, even if you find one that works, keep trying to see which one, which one, see which one, which strips work.

———

2. T: Three nine, # three nine, three nine, three nine three nine!
3. T: Three nine!
4. M: Thomas?
5. T: Three nine.
6. M: Three-ninths?
7. M: Thomas says three nine fits in there too.

———

8. T: Two-sixths works!
9. M: Good!
10. T: I know how it goes now!
11. M: How, how does it go Thomas?
12. T: Look, um, it goes by three, if the first number's three, it goes by three, and by one, it goes by one.
13. M: Oh, look at, everybody listen to Thomas.
14. A: hey, hey!
15. M: What are they called?
16. T: It's like um, if the first one is three it goes by three, and then uh, other three.
17. M: Where do use that, when it goes by threes?
18. T: Huh?
19. T: Times!
20. M: Okay so you're giving us some information here.
21. A: Sir, sir!
22. M: Thomas?
23. T: On day two, you messed up, two six, it's supposed to be two eight.
24. M: Oh yeah?

25. T: Yeah, it's supposed to be two eight.
26. T: It's going by um, times, um, two.
27. M: How do you call that, though?
28. T: Multiply!
29. M: Multiples.
30. T: Oh, multiples.
31. M: Okay, Thomas says they're multiples.

dent) was engaged and enthusiastic in the whole class lesson. Lines 2 and 3 show how Thomas is so enthusiastically engaged in the lesson, he uses the strategy of repeating and raising his voice to get the attention of his teacher. He is ultimately successful when the teacher gives him the floor (line 4, 5). Mr. Martinez then revoices, or repeats back Thomas' words, giving him credibility and highlighting his expertise and knowledge to the class. The teacher accomplishes this revoicing first with a confirming question, and then with a statement to the rest of the class (lines 6, 7). However, the teacher models the standard form in the first revoicing (line 6) but not in the second (line 7). Thomas' engagement within the participation framework with the teacher, playing the role of problem solver continues in the next segment. Lines 8 and 9 show how Thomas is recognized, validated, and praised by his teacher as a result of his response. Thomas has the confidence, as well as the autonomy to engage with the teacher and the rest of the class as he declares that he understands the pattern the fraction strips illustrate (line 10). Lines 11–19 show how not only Mr. Martinez participates in the framework Thomas has created with his declaration, but also how other students are invited—indeed—commanded to join the framework Thomas is leading (line 13). In line 20, Mr. Martinez' still includes Thomas personally in the discussion, but attempts to draw in other students by referring to himself as united with the other students (us). However, Thomas still wants to play the lead in the participation framework. He elicits his teacher's attention (line 21), is recognized (line 22), and then creates a new framework comprising the teacher and him. Here, Thomas shifts from being a problem solver to presenting himself as an expert who corrects the 'expert' (the teacher) (line 23). He continues this framework in lines 24 through 26, and Mr. Martinez agrees to be corrected. The teacher then redirects Thomas' focus (line 27). Mr. Martinez takes on the teacher role and Thomas becomes again the student. He encourages Thomas to think about a specific feature of mathematical discourse—multiples—and then scaffolds Thomas to reach the correct term (lines 28–30) by asking a focused question to draw out the answer (line 27). Mr. Martinez' closes the discussion, and the participation framework by revoicing, and ultimately giving credence to Thomas' contribution, in line 31. Lines 27–31 illustrates how the participation framework that develops helps Thomas verbalize the mathematical concept 'multiple' as well as produce the appropriate form.

This example shows how Thomas is encouraged and supported in his math class. He portrays himself as a math whiz kid, and the teacher recognizes him as such. He is allowed, and indeed, encouraged to play the role of the expert in math class. Thomas has the ability to work within the frame that CMP presents, and he appears to be motivated and certainly highly engaged while carrying out the tasks accompanying the frame—in this case, determining how much money the kids at Thurgood Marshall raised. This example also shows how Thomas sees the connections between the framed activity and the math concept. Declaring: "I think I know how it goes now!" (line 10) Thomas makes the connection between the mathematical concept and the frame that veiled it. Most of the other focal students were not able to make this connection by themselves. Although he did not have all the answers—as he still required the expertise of his teacher to help him find the right words, Thomas usually 'got' the mathematical concepts presented in class. He was thus able to interact often with his teacher as an expert. In contrast to Thomas, Benjamin was less likely to play the role of expert with his teacher.

Mr. Martinez, the carpenter-cum-math teacher, who was not a fan of CMP, did not often use the reform-based curriculum, so the frames his students were provided with usually hailed from their teacher's real-world experiences. Mr. Martinez posed thought-provoking situations where students would have to make sure the measurements for a house were appropriate. When drawing angles, one student had extended the line a bit too much. Mr. Martinez warned: "That little bit is gonna kill someone if you are an engineer." Other times, he would tell the students they would be financially responsible for carpentry errors. Once, after his students had incorrectly answered a question regarding multiples of 48, Mr. Martinez told his students: "That proves to me you are not carpenters, because you would know that you put studs at 16 inches on sheetrock."

In the example that follows, students are completing a warm-up activity in which they must calculate the number of inches as a fraction compared to a foot. They are converting fractions within a frame of carpentry. The task is to tell what fraction of a foot a given length is. As soon as Thomas understands the task, he is able to do the assignment and help his friends, Armando the fluent bilingual, and Benjamin, the ELL.

EXAMPLE 4.2
THOMAS HELPS BENJAMIN WITH CARPENTRY, JANUARY 28

T = Thomas
A = Armando
B = Benjamin

1. A: . . . I don't know how to do it.
2. T: You know a foot is twelve inches, right?
3. T: Make 'em into a fraction.

———

4. T: Six is half of twelve, three is one-fourth of twelve, one is one-twelfth of a foot, nine inches is three-fourth of a foot, eighteen inches is one and a half of a foot, four inches is one-third of a foot, two inches one-sixth of a foot, eight inches is two-third of a foot, fifteen inches is one and one-fourth of a foot, that's it!
5. T: What? (talking to himself)
6. T: That was too easy! (talking to himself)
7. B: How do you do this one?
8. T: Three inches is one-fourth!
9. B: One fourth?
10. T: Look, six, three is half of one-sixth, right?
11. B: yeah?
12. T: Of six, right?
13. B: One four is one.
14. T: One fourth.
15. B: Oh yeah.
16. T: Like, fourth, three times four is twelve, so a foot has twelve inches, so it's one-fourth.

As the clip begins, Armando is indirectly asking for Thomas' help. Thomas' identity as a helper thus emerges. In line 2, Thomas engages within the participant framework with a supposition and a tag question. Thomas assumes that Armando knows that there are twelve inches in a foot and confirms this with the tag, 'right?' Then, he explains that the measurements are to be expressed as a fraction (line 3). He then breaks away from Armando to do his work. The entire utterance reflected in line 4 shows Thomas summarizing the measurements to fractions (to himself), and in lines 5 and 6, he reflects on how easy the assignment is. When Benjamin engages in discursive participation within the carpentry frame, he, like Armando, elicits the help of Thomas (line 7). As evidenced by his response, Thomas agrees to help Benjamin (line 8). Lines 10–16 reveal the negotiation between the helper and the helped. But Thomas is not a gentle tutor. He is no-nonsense and expects his tutees to keep up with his explanations. Thomas prefaces his helping utterances with 'Look' and ends with 'right' (lines 10, 12). Thomas' tag questions do, however, encourage Benjamin to respond to his tutor's explanations.

Within the carpentry frame, Thomas was an engaged participant. He worked alone and as a helper and expert. In the participant frameworks created in this frame, Benjamin's roles changed insofar as he shifted from one needing help to one who 'got it'. Benjamin did not play the role of carpenter within this frame, but he did participate as Thomas' tutee. The carpentry frame provided Benjamin with an opportunity to interact and seek help. Within the frame, Benjamin was engaged in the participant framework with Thomas. He first engaged as one needing help. Throughout the negotiation, Benjamin admitted that he understood what his peer tutor was explaining to him. As Benjamin shifted from one needing help, to one who 'got it' and no longer needed help, he moved away from his 'I dunno' identity, toward an 'I know' identity.

While I concede the possibility that Benjamin may not have understood Thomas' assistance and agreed with him to save face, his work supported the evidence that he understood the carpentry assignment (he got a grade of 100). He also spent time working alone on the assignment following the tutoring shown in the clip. Based on his fallback refrain of 'I dunno' and because he often did poorly on in-class assignments, Benjamin was generally not successful in classroom endeavors. However, he did well on the carpentry assignment on this day.

The carpentry frame provided an opportunity for students to work together in small groups, something that Mr. Martinez did not often allow his students to do. According to Thomas, Mr. Martinez allocated class time for small groups "probably like two percent... or one point six (percent)." Thomas' identity as a 'math person' followed him outside of math class.

It should also be noted that since Benjamin had been in the U.S. for just two years, he was (somewhat) new to the English system of measurement. He, as well as others new to the U.S. might have had a particularly hard time learning mathematical concepts when having to translate through multiple systems of measurement. Nevertheless, carpentry was one of the most common frames of play employed by Mr. Martinez.

Mr. Martinez, like Ms. Loesely and Ms. Koch, also took advantage of opportunities to employ frames the students would easily identify with and be attracted to, such as food, sports, and money. Food can be used in a very versatile way to exemplify uses of fractions. Pizza is probably one of the more common, and more popular examples of fractions in math class. As a result, its mere mention usually piqued the interest of the majority of the sixth graders in class. Similarly, pies, cakes, and other desserts were often incorporated into class lessons to reveal some of the many ways that fractions are used in 'real life'. Within the frame of food, students played roles such as baker, cook, restaurant patron, or just an invited guest to a pizza lunch. Having the mathematical competency to make the right choice between a slice of pizza totaling either seven-eighths or seven-twelfths (see Chapter 3, Example 3.1, January 28) was quickly understood by all in the class to be an important and valuable skill in one's 'real life'.

When working with the fraction units at the beginning of the spring term, Mr. Martinez and Ms. Loesely incorporated many food-related frames in their lessons to help their students understand the relationship of part to whole and whole to part. In the next example, we turn to Ms. Loesely's class to understand how focal students Jennifer and Nestor learned within frames of play.

JENNIFER AND NESTOR—BROWNIES, PIZZA, AND FRACTIONS

In general, Jennifer was participatory in math class. She was not, however, extremely focused. In small group work, she could easily be swayed off-task, engaged in discussions in Spanish or English regarding friends or social activities. She admitted that she didn't like math that much, and agreed with her mom that math (language) was like 'Chinese writing'. Although she was less adept at getting correct answers, she could be motivated to engage in math discourse within frames of play.

In Example 4.3 from February 2, Ms. Loesely's third period class is working in the same CMP unit as Mr. Martinez' second period, *Bits and Pieces I* (Lappan et al., 2002a). The lesson is entitled "Cooking with Fractions" and students must invoke their higher-order thinking skills to analyze and evaluate the situation that first elicits student response regarding what size brownies should be made for the afternoon snack for 240 students attending school camp at Sturgis Middle School. It is also the same unit presented in Examples 3.7 and 3.7 of Chapter 3.

Since the class has been working within the frame of brownies for several days now, they are intensely involved in this topic. Students threatened each other with comments such as Jennifer's: "you don't get no brownies!"

which provided evidence that they were keenly aware they would actually be eating brownies in class when the unit concluded. This incentive served to motivate the students and also worked to integrate students within the curriculum and the frame of the lesson.

As Example 4.3 begins, the students and teacher discuss some of the ways in which a brownie pan can be divided depending on whether they make either small, medium, or large brownies. Part 2 of Example 4.3 occurs a few minutes later, in a partner activity which asks students to reflect on which size brownies they would make if certain conditions were in place (they would have to buy the ingredients, they would have to help make the brownies, or, they wouldn't have to do anything). Jennifer is working with her friend Kristina, but Kristina has left for the moment.

EXAMPLE 4.3
JENNIFER MAKES BROWNIES FOR 240 CAMPERS, FEBRUARY 2

L = Ms. Loesely
J = Jennifer
A = All

Part 1

1. L: Okay, six times five is thirty, okay, this is our large, these are our medium, this is the number of pans, and we take each pan, we need the medium brownies, we take our pans and cut 'em into twenty brownies like we did for this class, right?
2. L: You didn't get that big of brownie, did you?
3. J: No.
4. L: Cuz I made em this big, okay.
5. L: If we we're gonna make the small brownies and get thirty out of a pan, we'd only have to make # what?
6. J: Eight.
7. A: Eight.
8. L: Eight # pans.
9. L: Okay, if we we're gonna do large brownies, we could only get fifteen out of a pan, how many pans would we have to make?
10. J: Sixteen.
11. L: Okay.

Part 2 (Several minutes later, the same day)

12. J: Miss # Miss, I didn't get number, we didn't get number two.
13. L: Number two?

14. J: You have to make the brownies.
15. L: What if you were the one that was in charge of standing there mixing everything up and making the brownies, what size would you wanna make?
16. J: The smaller one so you get done faster, you get done faster, cuz like what if you have plans, and like you just wanted to get done fast?
17. J: But to be fair, just make the medium ones.
18. L: Okay, anybody not ready to discuss?

These examples show how Jennifer engaged within the brownie frame. Her participation in the discursive interchange between teacher and student shows that she acknowledges brownies are used to discuss math and proves that she comprehends all of the teacher's questions. In part 1, Jennifer responds appropriately to her teacher's questions (lines 2–10). While Ms. Loesely engages within a participation framework with Jennifer, she maintains her role as expert, and Jennifer plays the role of the student, with the right answer.

In part 2, Jennifer continues to be motivated and engaged in the brownie frame. She supplies thought-provoking responses to the hypothetical questions about baking brownies for 240 students and displays sophisticated higher order thinking. The participant framework that is created in part 2 is different from the teacher/student framework from part 1 which is much more personal and dynamic. When Jennifer elicits the assistance of her teacher (line 12), Ms. Loesely counters with a clarification (line 13). With her teacher's full attention, Jennifer then reminds the teacher of question 2, the one she needs help with. The complete utterance reflected in line 15 shows how the teacher incorporates a different frame of play within the brownie frame, to help Jennifer reach her conclusion. Ms. Loesely attempts to reify the act of cooking by providing an unpleasant scenario, "what if you have to cook..." prefaced with the framing introducer, 'what if' (line 15). Jennifer acknowledges and buys into her teacher's frame, and then comes up with a response. Her answer, interestingly, is manifested in a format similar to Ms. Loesely's; it also includes a 'what if' introducer and follows with a possible situation (line 16). Jennifer supplies the possible situation of "what if I have plans" to assist her in making a decision regarding the size of brownies she'll make. Finally, though, in line 17, she concedes that it would likely be the fairest possible choice to meet in the middle and make the medium-sized brownies. With that comment, Jennifer's teacher officially breaks up the participation framework between the two of them, and attempts to corral the attentions of the other students in the class (line 18) so they can commence a full-class, teacher-moderated discussion.

In Example 4.2, Jennifer displayed higher-order thinking and rational, thoughtful responses to the thought-provoking questions presented in the text. In both Parts 1 and 2 she was engaged and motivated by the brownie problem. Jennifer's mathematical discourse was encouraged through teacher scaffolding while interacting with her in a participation framework. Interestingly, Jennifer employed similar discursive strategies as her teacher within the frameworks. Jennifer's tentative discourse marked by *if* and *what if* followed a similar format of her teacher's. As a result of the conditional responses that Jennifer supplies, it is evident that she has been socialized to accept certain frames and the presuppositions they encompass. Her use of 'teacher' discourse also indicates that she recognizes not only that Ms. Loesely uses a variety different from hers, but also that this teacher variety belongs within the realm of mathematical discourse available for use by students. Having the opportunity to engage within participant frameworks with her teacher, Jennifer was able to emulate her model (the teacher), in one aspect of mathematical discourse. Although Jennifer was less successful when carrying out mathematical tasks alone, she proved she could think creatively and use mathematical discourse appropriately within lessons framed as play. The social setting which the frames of play provided invited a variety of participation frameworks in which Jennifer and her classmates could interact. Nevertheless, the scaffolding provided by Ms. Loesely was critical to Jennifer's success on the mathematical responses within the frame. Next we will see how Nestor participated within frames of play in Ms. Loesely's class.

On February 4, Ms. Loesely's third period are doing ACE (Applications-Connections-Extention) questions from *Bits and Pieces I.* The students are working in partners on a problem that asks "How many fourths are in four and a fourth?" (Lappan et al., 2002a, p.35). In Example 4.4 Nestor is working with Jon, a semi-engaged student who normally made pretty good grades in Ms. Loesely's class.

EXAMPLE 4.4
NESTOR—THINK ABOUT PIZZA, FEBRUARY 4

L = Ms. Loesely
N = Nestor
J = Jon
A = All
B = Boy

1. J: How many fourths are in four and one fourth?
2. N: Which one?

3. J: Four, oh, four?
4. J: Sixteen?
5. N: How many fours are in four?
6. N: Two, (laughs), you know it?
7. J: Four, okay four.
   (Some students say they don't get it).
8. L: Think about pizzas, how many whole pieces do we have?
9. N: Four.
10. A: Four!
11. B: What's fourth of one?
12. T: Okay and we have one-fourth of one more, well how do we know how many fourths are in four?
13. J: I dunno.
14. N: Four.
15. L: Divide them into?
16. B: Two.
17. L: Two?
18. A: Four, four.
19. L: So, Nestor, count the fours, and tell me how many fours we have.
20. N: Hmm.
21. L: Four and one fourth.
22. N: Um.
23. L: You can't count 'em down there Nestor, it's up here on the overhead.
24. N: Sixteen?
25. N: One, two, # one, two, three, four, five, six, seven, eight, nine, ten, eleven, twelve, thirteen, fourteen, fifteen, sixteen, seventeen.

Lines 1–7 show a participant framework between Nestor and his partner, Jon, where they are students and equals. At first Jon enlists help (line 1), and Nestor restates the question (line 5) to understand it (Nestor was faithful to the learning strategy of restate, plan, work, and check, he admitted in an interview with me. He laughs when he quickly realizes he knows the answer to Jon's question (line 6) and uses his knowledge to encourage his partner. Within the same participant framework, Nestor played the role of equal and helper. When someone requests assistance in understanding the problem, Ms. Loesely introduces the frame of pizza (line 8). Although she does not continue referring to pizza, she implies it in the complete utterance in line 12 when she asks "how do we know how many fourths are in four?" Nestor quietly responds, but not with the answer she wants (line 14). Lines 19–25 show Ms. Loesely engaging with Nestor to count the fours. He is admonished for looking at his book instead of the overhead, where he should have been looking. In the participation framework he engages in with his teacher, Nestor plays a sheepish, off-task sleepyhead.

In the previous example, the introduction of the pizza frame as a thinking strategy was not very effective. While Nestor and others eventually understood the problem, the pizza frame did not help them. It was clear that in order for the frame to have been more effective for Nestor and his classmates, Ms. Loesely needed to draw upon it more heavily within the discussion, but we will return to possible effective techniques at the conclusion of the chapter. The following section illustrates how Ms. Koch used frames with the learners in her mathematics classes.

ANNA AND PEDRO—MONEY, ROLLER DERBY, AND PROBABILITY

In the same way that the teachers used food to interest and connect with students learning fractions, money is often employed by both teachers and students to illustrate use of fractions and decimals. Roles played by mathematics students when working within the frame of money included serving as a customer, a waiter, or a banker. Money was used quite often to encourage students to think about mathematical concepts and to provide a beneficial learning strategy for them to better understand the concepts. In the following examples from Ms. Koch's class, she invites her class to think about money in order to understand how fractions can be divided.

EXAMPLE 4.5: ANNA AND FOUR QUARTERS, DIVIDING FRACTIONS, FEBRUARY 5

K = Ms. Koch
A = Anna
B = Boy
AL = All

1. K: How many times would twenty-five go into a hundred and twenty?

———

2. A: Four. (Self-talk)
3. K: What do we know about money?

4. B: Four.
5. A: Four quarters.
6. K: Let's count, now, think about quarters, okay, twenty-five, fifty, seventy-five, one hundred, so how many times is it going to go?
7. AL: Four.

Ms. Koch has presented the class with answering the problem of 25 divided by 120. Shy Anna is answering the teacher, but under her breath (line 2). When Ms. Koch understands that the class needs a hint to motivate the response, she invokes their understanding of money in line 3. In this teacher-fronted whole class participation framework, even Anna becomes involved upon Ms. Koch's introduction of the money frame. Anna takes Ms. Koch's cue and provides the appropriate response in line 5, four quarters. The teacher then uses the frame of money to involve the students in the question, and the answer, four, is achieved, in line 7. Anna is included in the group choral response of line 7.

Just six class days later, Pedro and his group mates, Sarina, Anna, and Emily were counting up how many times each of them had conducted a trial involving the selection of colored blocks from a bucket. They were expected to conduct 25 trials each and then multiply that number by the number of people at their table, in this case, 4. As a group, they would end up with a fraction for each of the colors that had been pulled out of the bucket, and were to total each of the tallies up of each color.

EXAMPLE 4.6
PEDRO USES THE MONEY FRAME WITH PROBABILITY, FEBRUARY 13

P = Pedro
S = Sarina
K = Ms. Koch

1. P: How many?
2. S: Twenty-five, twenty-five, twenty-five,
3. S: twenty-five, that's how many.

———

4. P: It's twenty-five cents, twenty-five cents, twenty-five cents.

———

5. P: What if I just count 'em all?
6. K: We'll be here all day then.

In lines 1, 2, 3, and 4, Pedro and Sarina are counting their individual numbers of tallies, but Pedro invokes the frame of money in line 4. He does not continue with this example beyond his single utterance, but seems to be using money to help him think about how he will calculate the total number of tallies for the group. In line 5 he proposes that he simply count up all of the tallies for each color to get totals, however, Ms, Koch provides a clear (but nonetheless indirect) indication that his suggestion is unacceptable. Pedro's implementation of a money frame suggests that he is motivated to find an acceptable answer to the teacher's question, and that the frame may help him to find that answer. In effect, the use of money may be the learning strategy for Pedro that Ms. Koch had originally intended it to be. The last example of frames as games comes from the CMP unit, *How Likely Is It?* and shows Anna engaged in participation frameworks with other students, while learning about gaming and probability.

In January and February, Ms. Koch introduced the concept of probability to her third period. In the CMP text *How Likely is it?* students were presented first with concepts relating to experimental probability and then theoretical probability. The activities in the chapter on experimental probability are very hands-on and require a good deal of trial and error experiments involving spinners, dice and coin tosses. Ms. Koch attempted to frame a lesson on February 10 by introducing it with this enticing statement: "We're going to be analyzing games of chance today so it's almost like going to Las Vegas today!" But the students' silence suggested that they were not moved by this enthusiastic introduction. As the activity ensued, however, the students became involved, participatory, and enthusiastic. The activity they conducted required students to tally and analyze the possibilities of number combinations that arose from the tossing of two dice in a game called Roller Derby. Anna was an enthusiastic participant in her small group on the day they played Roller Derby. Anna's small group consisted of Sarina, a bright, talkative girl, and Emily, who according to Anna, was "mean to (her)" in the past.

EXAMPLE 4.7
KOCH, ROLLER DERBY, FEBRUARY 10

A = Anna
E = Emily

1. A: No, oh, I got, now, five now, five, seven, no seven, Emily, Sarina.
2. A: *Él no es,* six, no, ay, six, I need two [it isn't, no, ay, six, I need two]!
   —— (rolls again)

3. A: Nine ## three six, nine.
4. E: That's ten!
5. A: Oh, you have more, you are mean!
6. A: Six, go!
7. A: What, but it was eleven, it wasn't eleven, Emily move it.
8. A: She wanna [: wants a] one.
9. E: Eleven.
10. A: No, it's not eleven ## oh yeah, eleven, you won . . . oh, it's bad!

Although shy Anna often took a back seat to her group mates, allowing them to lead the interaction during small group work, today she decided that she was willing and enthusiastic to participate, and even to take a major role in the roller derby game. Line 1 shows that Anna is directly addressing two of her partners (that she was not fond of and had described in a one-on-one interview as 'mean'—due to the fact that Emily had made fun of Anna's language ability publicly). By inviting her group mates into her discursive framework, Anna creates a participation framework in which she serves as an equal, and perhaps even dominant partner. The next two lines (2, 3) show Anna simply describing her roll. She has apparently made a mistake in the reporting of her roll, as evidenced in Emily's overt correction in line 4. Anna could have easily shifted into a subordinate role by allowing Emily to silence her in the stern, teacher-like tone Emily icily uses, but Anna is not fazed by Emily's comment. We see in line 5 that Anna evaluates Emily's number cube roll and feels comfortable enough to pass a somewhat negative, but seemingly quite playful value judgment ("you are mean" line 5) on her self-proclaimed nemesis. Anna continues to evaluate the number cube score and at the same time, directs Emily to take her next roll (lines 6, 7). Anna then reveals that she knows what Emily needs to win (line 8). Anna's analysis of Emily's game playing continues in line 10 when Anna assesses her opponent's final tally, and realizes that she has lost.

Within this short example, Anna has proven that she has the linguistic and mathematical ability to use a variety of discourse types. She is able to give directions and encourage her peers to carry out their required tasks, she can analyze and assess rolls on a number cube, give simple descriptions of the stage of the game. She can determine who lacks what information (in this case, numbers) to achieve success, and can evaluate the winner and loser of the somewhat complicated mathematical game of probability. Normally, Anna was too reserved to reveal her ability in both English and mathematical game playing, but the enthusiasm for the game and the excitement that the activity brought on helped her tremendously to feel comfortable, or in Krashen's (1982) terms, to "lower (her) affective filter" and use her second language.

The participation frameworks that Anna helped create during the game put her in an equal and even dominant role in her small group. As a result of her enthusiasm, interest, and her ability to succeed in the classroom task, Anna felt comfortable to engage in and even direct the mathematical discourse within her small group activity. The activity of the day, which had been framed as a game, played a major role in Anna's discursive interaction within the participant frameworks ongoing in the day's lesson.

The frame of the game worked to illustrate the principles of probability for Anna and her class. As an ELL, the game frame, as well as the others that were commonly employed in the CMP-inspired sixth grade math classes, was highly beneficial because the activity which grew out of the frame was interesting and motivating, and it allowed the ELLs to become involved in the task. In the roller derby game, Anna was an active participant in not just the small group portion of the class, but also the teacher fronted discussion part of the lesson. Although she was still somewhat quiet, Anna answered the teacher's questions with yes, no, and numbers which were directed at her regarding outcomes from the game. Anna's first-hand experience with the game made her better able to relate to the teacher's questions regarding the game and thus better able to successfully interact in whole class interactions.

Frames, or interesting, motivating scenarios, designed to help learners understand mathematical concepts, are used throughout the CMP curriculum. For ELLs and bilinguals, the frames assist learners to participate in mathematical discourse and interaction by piquing their interest, making math fun, providing thought-provoking problems, and encouraging discursive interaction with others. In addition to the frames introduced in the CMP classroom, frames provided by the teachers helped to engage ELLs in mathematical discourse.

In general, frames of play helped the six focal students in this study to understand how fractions and probability were used in context. However, the focal students who were less successful in mathematical endeavors, such as Nestor and Benjamin, and to a somewhat lesser degree, Jennifer did not reap the benefits from frames of play that the others did. In some cases, Nestor, Benjamin, and Jennifer were not able to see the connections between the mathematical concepts and the play frames in which the concepts were contextualized. Differences in the amount and manner in which each of their teachers utilized play frames contributed to these students' learning and success, as did the amount of group work, teacher scaffolding, and individual student skill level in math. Next I will delineate these limitations of learning within play frames and participation frameworks by comparing and contrasting focal students from the three different classes.

COMPARISON OF FOCAL STUDENTS

Here I contrast the learning of focal students through frames of play and participant frameworks which emerged in classroom activities. I have chosen to compare and contrast two focal students from different classes, based on their background (where born, time in U.S. schools) and their classification as a language learner, based on their current or past language support in school. The students I have matched up are shown in Table 4.4.

Thomas and Jennifer: U.S. Born, Exited from Language Support

In comparing the relative success of Thomas and Jennifer, their participation within frames should be measured with respect to the use of frames and emergence of various roles within participant frameworks. Based on Thomas' consistent and steady engagement in whole class mathematical discussions, it is evident that the classroom environment in which he was placed worked within his learning style. Mr. Martinez did not use a great deal of frames of play, but it is not clear that Thomas reaped as much benefit from them as others did. When it was necessary (when Mr. Martinez framed the lessons) Thomas would work within the parameters of the frame, but he was just as engaged in lessons that were nothing more than straight calculations. One thing that Thomas did benefit from within framed lessons was interaction with the teacher. Mr. Martinez' traditional teacher-fronted class allowed Thomas to play the role of not only math whiz kid, but also expert, teacher, and mathematician within participant frameworks created with the teacher, and occasionally with his classmates. He would not have had such autonomy nor 'floor time' to speak to the teacher and the whole class in a more group-centered classroom. Thomas also demonstrated that he had

TABLE 4.4 Comparison of Focal Students from Different Classes

Classification	Teacher	Student	Exit/entrance
U.S. born, exited from language support:	Martinez	Thomas	(exited in 3rd grade)
	Loesely	Jennifer	(exited in 6th grade)
U.S. born, currently receiving ESL services:	Loesely	Nestor	(bilingual/ESL throughout elementary)
	Koch	Pedro	(bilingual/ESL throughout elementary)
Mexican born, recent arrivals, currently receiving ESL services (since U.S. arrival):	Koch	Anna	(arrival in 4th grade)
	Martinez	Benjamin	(arrival in 4th grade)

the skill and the competence to 'see through' the frames encompassing the mathematical concepts. He was one of the few students who could get the mathematical point veiled within a frame of play.

Jennifer benefitted from the use of play frames. When working in small groups, she was engaged and motivated to complete the lesson. The participant frameworks that developed for her allowed her to play both the expert and the novice, leader and follower. With respect to the frames imposed by her teacher, Jennifer clearly understood that she needed to work within the parameters of the frame (as she did when making brownies for 240 students). She was able, and in fact, rather highly skilled at discussing the open-ended problems that her CMP text posed. She demonstrated that she could utilize the kind of tentative discourse employed in hypothetical situations—something she modeled after her teacher. However, when time came to find solutions with one 'right' answer, she was less skilled. What Jennifer needed more of was a dependable mentor and more scaffolding to help her understand and check her calculations and solutions. The one on one relationship that Thomas shared with his teacher, Mr. Martinez' would be good for someone like Jennifer. Another difference between Jennifer and Thomas was that Thomas was able to see through the frames which contextualized the math problems, but Jennifer was not. When Thomas stated: "I know how it goes!" in Example 4.1 of this chapter, he demonstrated his understanding of the math concept of the day. In contrast, when I asked Jennifer what she had been learning in math class lately, she replied: "Brownies!" While learning math within the frame of brownies, Jennifer did not see the connections between the math and the context, as Thomas so easily did.

Pedro and Nestor: U.S. Born, Currently Receiving ESL Services

Both Pedro's and Nestor's teacher utilized frames of play in their math classes. Pedro the go-getter could distinguish between the frame and the math, as Thomas could, but it is not clear that Nestor had that capability. Pedro's teacher, Ms. Koch, explained that she filled in the gaps that CMP left behind with additional resources such as warm-up activities, other textbooks, and worksheets. If she believed the students had problems making critical connections between the CMP lessons and the mathematical concepts, she would address the issues with mini-lessons. Furthermore, Pedro and his partners, usually comprising Anna, Emily, and Sarina, were always seated as a group, and often were expected to solve a problem, play a game, or conduct an experiment as related to the particular frame the class was learning. Pedro was thus provided with opportunities in which to practice, manipulate, and discuss the mathematical concepts presented in the CMP text. He could play various learning roles while using play frames and working within participant frameworks. Example 4.5 illustrated how Pedro used a frame commonly used in his classes in both whole class discussions and small class interactions in much the same way that his teacher did. By using the frame that he did, Pedro showed that he understood the utility of how frames are used as a learning strategy. He also demonstrated socialization to and the use of frames in an academic setting. As a result, Pedro had the opportunity to make connections, with the help of his teacher.

Since Nestor's teacher, Ms. Loesely, was not 'allowed' to conduct warm-ups (as mandated by the school administration), she did not feel she had time to introduce new material not directly related to the state standardized test. Unlike Ms. Koch, whose students worked in teams and had completed several CMP units in advisory class, Ms. Loesely was not able to complete all sixth grade CMP lessons. She and her colleague, Mr. Martinez, could only pick and choose the CMP lessons they felt would be most beneficial to their students. As many of the frames used in class came from the CMP units, and Ms. Loesely used CMP less than Ms. Koch, Ms. Loesely's students participated in fewer than Ms. Koch's students.

Similarly, in Ms. Loesely's class, students such as Nestor had fewer opportunities to engage in small group work. Although Ms. Loesely did provide for group work on particular days, Nestor and his classmates had less of an opportunity to create a variety of participant frameworks than Pedro and his classmates did. Therefore, Nestor did not receive the opportunities for engagement within participant frameworks or the amount of interaction with motivating frames that Pedro did in his class.

Anna and Benjamin: Born in Mexico, Recent Arrivals, Currently Receiving ESL Services

Anna and Benjamin both benefitted from the frames of play. Frames of play helped Anna to learn in several ways. When learning fractions and probability within frames of play, she was often more motivated, enthusiastic, extroverted, and as a result, more participatory. Because Anna had more extensive opportunities to create various participant frameworks in class, she gained more experience playing different roles within them. On some days, she was a leader in small group activities; on others, she was a helper to her classmates. Working within the small group activities that frames of play provided, Anna was able to engage in mathematical discussion and interaction. In whole class activities, Anna was usually reserved and shy. Within smaller activities, she opened up and participated richly. In Ms. Koch's class, Anna had more opportunities to learn math within frames.

Benjamin did not have the same opportunities that Anna did. Fraction lessons for Anna were often framed within interesting scenarios centering around restaurants, games, candy, and cereal. While some of the fraction lessons that Benjamin studied in class were also framed within interesting scenarios, many were not presented within a framed scenario. Benjamin did not have the advantage that Anna had in learning within the quantity and variety of play frames. Another difference between Anna and Benjamin's learning situations was that Anna worked in a variety of grouping structures. Benjamin's class did not do much small group work. Small group activities often open up a variety of participation frameworks within which learners can interact and learn together. As a result, Benjamin had fewer opportunities to engage in participation frameworks than Anna did in her class. Benjamin may have benefitted from more engagement in small group configurations, unfortunately, these opportunities were not provided in Mr. Martinez' class.

In the next chapter, we will take a further look at how each of these students talked in and about mathematical discourse and see how each was successful with respect to using the discourse and doing math.

NOTES

1. Each of these concepts is discussed within the literature on discourse analysis. See for example Tannen (1993) or Schiffrin, Tannen, and Hamilton (2001).
2. There are 8 units in each textbook series to be taught per year.

CHAPTER 5

WHAT IS MATHEMATICAL DISCOURSE, HOW IS IT USED, AND WHO IS SUCCESSFUL AT IT?

In this chapter, I address key issues related to mathematical discourse and its development by ELLs. First, I explicate how researchers, educators, and the CMP curriculum define mathematical discourse, specifically for the six sixth grade focal students highlighted in this study. I then outline how that discourse is manifested within one sixth grade CMP fraction unit. I then describe the discourse practices used by the six focal students and address how these discourse practices contribute to success in mathematics through central or peripheral participation. Last, I present my explanation for the process of oral discourse development. I have outlined three types of participation that I call participation performances. These participation performances include passive/receptive learning, experimentation, and ownership. I illustrate these performances with examples of focal students' oral discourse and draw conclusions regarding these performances and participation in the classroom CoP.

English Language Learners and Math, pages 95–120

WHAT IS THE MATH DISCOURSE WHEN LEARNING FRACTIONS?

Each subject area has its own particular discourse which entails (discourse) patterns, syntax and semantics, a specific lexicon, and certainly, the background knowledge inherent in the discipline itself, which is requisite for successful comprehension of situated content language. Attempts have been made to characterize the academic discourse of specific content areas, but this is no simple task. There are a multitude of facets of each and every academic discipline that make categorizing and classifying all the features of one particular academic discourse a near impossible venture. It is no less difficult in mathematics. In the content area of math, Brilliant-Mills argued that a specific mathematical discourse does not exist. Instead, she maintained that there are a variety of 'situationally constituted' math registers or discourses, which develop through patterns of interaction (Brilliant-Mills, 1994, p. 302).

As noted earlier, the National Council of Teachers of Mathematics (NCTM) defines the specialized, situated, and contextual language of math as "ways of representing, thinking, talking, agreeing, and disagreeing" (1991, p. 34). The organization (representing mathematics professionals) describes math discourse in a somewhat general and all-encompassing manner. Moreover, NCTM, in its *Principles and Standards for School Mathematics* (2000) separates principles from standards, and further delineates content from process standards. Within the process standards, under the heading of communication, we understand NCTM's position on mathematical discourse. According to NCTM, "communication is an essential part of mathematics and mathematics education" (2000, p. 60). It is the goal of NCTM that all pre K–12 grade students should be able to:

- Organize and consolidate their mathematical thinking through communication;
- Communicate their mathematical thinking coherently and clearly to peers, teachers, and others;
- Analyze and evaluate the mathematical thinking and strategies of others;
- Use the language of mathematics to express mathematical ideas precisely (NCTM, 2000, p. 60).

The communication guidelines proposed by NCTM do not refer to any particular aspect of math content (such as fractions, prime numbers, geometry, or algebra). Therefore, a specific, detailed definition of mathematical discourse regarding fractions is not readily supplied in the literature. The CMP curriculum, however, provides some specifics regarding aspects of what math discourse is. The focus in the texts was, for the most part, related to vocabulary.

CMP

To assist me in answering the question regarding the definition of expected mathematical discourse knowledge, I examined the CMP curriculum. In particular, I looked at the three CMP texts used to teach fractions (and fraction-related concepts such as probability and ratios) in sixth grade: *Bits and Pieces I, Bits and Pieces II,* and *How Likely is it?* I first examined the glossaries of each of these texts and found critical mathematical concepts for each book's lessons. Within these explanations for mathematical concepts were important vocabulary and terminology. The terms highlighted in the glossary were given more than adequate treatment. Each term was described and defined in length (sometimes in more than two paragraphs), and often was followed by an illustrative diagram to exemplify the lexical item listed. In *Bits and Pieces I* and *II,* many of the same terms were included in each of the glossaries. One all- important vocabulary word and mathematical concept discussed in these books is *fraction.* This term is defined with the following explanation: A number (a quantity) of the form a/b where a and b are whole numbers.[1] A fraction can indicate a part of a whole object or set, a ratio of two quantities, or a division. For the picture in Figure 5.1, the fraction 3/4 shows the part of the rectangle that is shaded: the denominator 4 indicates the number of equal-size pieces, and the numerator 3 indicates the number of pieces that are shaded.

The fraction 3/4 could also represent three of a group of four items meeting a particular criteria; the ratio 3 to 4 (for example when 12 students enjoyed a particular activity and 16 students did not); or the amount of pizza each person receives when three pizzas are shared equally among four people, which would be 3 ÷ 4 or 3/4 of a pizza per person (diagram of three pizzas follows in original) (Lappan et al., 2002b, p. 93).

Thus, the explanations and definitions for each concept included in the glossary are quite detailed in terms of their syntactic structure and illustrative nature. They define and describe vocabulary, but they also illustrate

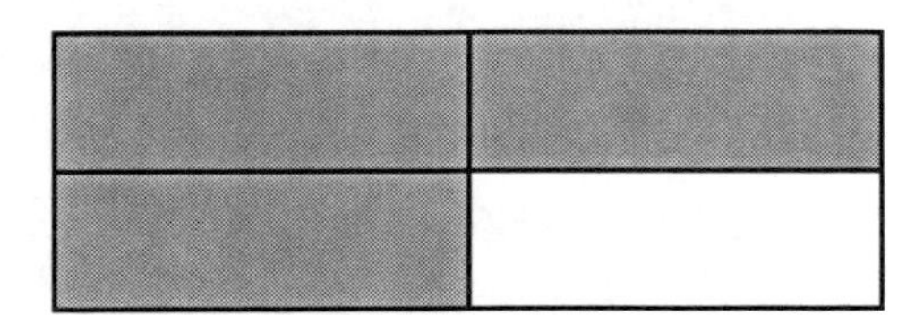

Figure 5.1 CMP Fraction (Lappan, et. al, 2002b, p. 93).

how the concept is used in context. The glossary definition for *fraction* encompasses further vocabulary words critical to the learning of fractions such as *number, quantity, part, set, ratio,* and *division.* These terms have specific mathematical meanings. A language learner may understand *part* and *set,* for example, but they may not know that these terms have discipline specific meanings in math class. Further, a language learner may have difficulty comprehending the seemingly simple explanation that "A fraction can indicate a part of a whole object or set, a ratio of two quantities, or a division." The difficulty may arise out of lack of lexicon, lack of context, or a misunderstanding of semantics. The second part of the definition introduces additional abstract notions (*. . . could also represent three of a group of four items meeting a particular criteria*), complex syntax (*the ratio 3 to 4 (for example when 12 students enjoyed a particular activity and 16 students did not), or the amount of pizza each person receives when three pizzas are shared equally among four people)* and math specific symbols *(3 ÷ 4 or 3/4 of a pizza per person.)*

Although there are relatively few chosen lexical items included, they are given thorough treatment. The list in Table 5.1 details the important terms included in the glossaries of *Bits and Pieces* and *How Likely Is It?*

While the list includes primarily lexical items and short phrases critical to the understanding and use of fractions and fraction-related concepts, the definitions in the glossary and in the individual units of the CMP texts both promote knowing of how to utilize and manipulate these concepts discur-

TABLE 5.1 Fractional Concepts: Vocabulary and Complex Phrases from CMP Texts

Bits and Pieces I	*Bits and Pieces II*	*How Likely is it?*
	algorithm	certain event
*base ten number system	*base ten number system	chances
*benchmark	*benchmark	equally likely events
*decimal	*decimal	event
*denominator	*denominator	experimental probability
*equivalent fractions	*equivalent fractions	fair game
*fraction	*fraction	favorable outcome
*numerator	*numerator	impossible event
*percent	*percent	outcome
	reciprocal	possible
*unit fraction	*unit fraction	probability
		probable
	random events	theoretical possibility
		trial

* included in books I and II

sively in an experiment or exercise. For example, when students in Ms. Losely's and Mr. Martinez' class were conducting the paper-cutting fraction strip activity at the beginning of the spring term, students needed to understand how to follow directions, specifically in command/imperative form, and to comprehend and produce expressions such as 'mark the folds', 'fold it lengthwise', 'use your cross products', strips, fractional parts, and representations of different fractions. Therefore, in each individual unit, ELLs and native English speakers alike are asked to demonstrate competence in a variety of productive and receptive linguistic tasks. The benefit of CMP for ELLs is that in many of the classroom situations, such as in the fraction strip lesson, instruction is facilitated by interactive tasks and hands-on activities.

On page four of each of the fraction-related CMP books, a passage entitled 'Mathematical Highlights' appears. The introductory sentence in *Bits and Pieces II* tells students the primary learning objectives they will cover in the book, namely: "In *Bits and Pieces II*, you will develop understanding of algorithms for operations with fractions, decimals, and percents" (Lappan et al., 2002b). After which, several (seven or eight) bulleted examples describe the individual learning tasks the student will be expected to know as a result of working with the text. For example, *Bits and Pieces II* should help (students) to: "Use benchmarks and other strategies to estimate sums, differences, products and quotients" and "Develop ways to model sums, differences, products and quotients, including strip models, number line models, and area models" (Lappan et al., 2002b). All of these sentences are constructed with active and passive tenses of verbs such as 'model', 'analyze' 'understand', 'use', 'look for', 'compare', 'experiment with', 'become fluent' and 'move flexibly' (Lappan et al., 1998; 2002a,b). In some cases, a student would be able to demonstrate their understanding of the aforementioned concepts with a simple mathematical calculation (devoid of oral or written discourse), but most activities presented in the texts (to practice the concepts) are group-oriented, language-based tasks. In other words, ELLs and native English speakers must carry out CMP's prescribed mathematical tasks using oral and/or written language.

Each 'Mathematical Highlights' section follows with a tip or suggestion to students to think on their own as mathematicians when working with the new concepts presented in the book. *How Likely Is It?* tells students they will decide when calculators will be a useful tool to help them solve problems (Lappan et al., 1998). *Bits and Pieces I* and *II* direct students to ask questions when reading and working with the concepts presented in the book such as "What models or diagrams might be helpful in understanding the situation and the relationships among the quantities (in the problem)?" (Lappan et al., 1998; 2002b, p. 4). Discursive features related to thinking critically, analyzing, estimating, and implementing strategies when understanding fractions and finding answers to fraction problems are expected of students. However, these language-based tools may or may not have been pre-taught in earlier lessons. In CMP lessons, students learn mathematical discourse inductively. Students are presented with mathematical concepts and problems with the goal being to develop critical thinking skills through use of precise mathematical discourse.

In essence, Brilliant-Mills (1994) is correct. There is no one definable discourse in math; there are many different discourses in math. The lessons, curriculum, state and national standards, school district, the teachers, and even the student interactions serve to create and define the mathematical discourse used by sixth graders when studying fractions. There are, however, critical concepts, vocabulary, purposes for language, and ways of talking that are used to interact within academic groups that sixth graders must begin to utilize and negotiate with in class. The target form of discourse that ELLs, bilinguals, and native speakers alike are expected to learn in middle school math classes in the U.S. is English mathematical discourse as outlined in state essential skills guidelines, state standardized tests, and the CMP texts. By interacting with each other in a classroom setting, students learn how to use math discourse and think critically; in other words, they learn concepts and ways in which to express them in school. In the following section, we will examine how the six students used this somewhat ambiguous mathematical discourse in a CMP lesson in class.

HOW DO STUDENTS USE MATHEMATICAL DISCOURSE?

When examining student discourse within my classroom language data, I attended to two primary constructs present in the ELLs' school talk: language functions and higher order thinking skills. I chose to examine the presence and degree of these particular constructs due to their necessity in school communication for both interpersonal interactions in academic and social tasks in school and for academic written tasks, such as homework, tests, or writing assignments.

Language functions are the communicative functions involved in an oral, interactive, discursive interchange. Halliday (1978) outlined seven critical functions of language: *instrumental* (getting needs met), *regulatory* (controlling others' behavior), *informative* (communicating information), *interactional* (establishing social relationships), *personal* (expressing individuality), *heuristic* (investigating and acquiring knowledge), and *imaginative* (expressing fantasy or possibility). Peregoy and Boyle (2008) describe the use of some of these functions in the classroom. They note that:

> ...as teachers, we are charged with the task of developing students' abilities to use language effectively for heuristic and informational functions. Thus

> K–12 schooling serves to expand children's and young adults' repertoires in a variety of oral language functions. (pp. 120–121)

Diaz-Rico (2008) also maintains that "providing English learners with opportunities to engage in the various functions is critical for enabling them to develop a full range of proficiency in English" (p. 30). Thus, while language functions are a key part of the academic language students are expected to know to be successful in school, it is necessary that ELLs have sufficient opportunities to utilize and practice using these functions.

With respect to academic language, there are a plethora of language functions that are used in classrooms every day. Gibbons (1993) names a few: agreeing and disagreeing, apologizing, asking for assistance or directions, classifying, commanding/giving instructions, comparing, criticizing, describing, evaluating, explaining, among a multitude of others. Pinnell (1985) adapted Halliday's classification system for language functions to represent academic functions of language. For example, within the instrumental function, students would ask for assistance or directions, or within the informative function, students would command or give instructions, or even explain a notion they had learned in math or science class.

Within these functions, however, there is another dimension of language that is often stressed in school. That is the use of *higher order* forms of language. Bloom et al.'s (1956)[2] seminal classification system outlines knowledge from lower to higher as *knowledge, comprehension, application, analysis, synthesis,* and *evaluation,* and is used commonly in education to assess and determine complexity of academic language and thinking. This system, then, facilitates our identification of academic language, and can also help educators pinpoint where students need additional assistance in using more complex features of academic language functions.

In my examination of the six ELLs' language, I chose to highlight their use of academic language functions and higher order thinking skills in language. In particular, I analyzed the students' diversity of use of language functions, and the degree to which focal student language reflected use of complex higher order language. In my analysis, lower order language functions included simple counting, describing, confirming, revoicing another's words, and giving simple directions; whereas higher order language functions included giving reason or rationalizing, evaluating, questioning, and verbalizing complex calculations. This analysis helped me to understand how some students were ultimately successful in math (according to traditional measures of success such as grades, test scores, and participation in group activities).

This analysis also revealed how three of the focal students emerged as central participants in the classroom CoP, and how the other three did not meet the criteria outlined for centrality. Lave (1991) maintains that the shift from legitimate peripheral participation to central, or full participation involves "developing knowledgably skilled identities" (p. 65). That is, community members' identities must shift and transform from apprentice or novice to expert as they share mutual engagement in a joint enterprise, and exhibit proficiency in the shared repertoire of the group. Lave and Wenger (1991) maintain that the shift to centrality requires a participant to "make the culture of practice theirs" (p. 95). When one's identity transforms to the point of having made the practice his or hers, and is skilled at interacting within the multifaceted constructs of that particular practice, then the participant has attained the status of a full, or central participant.

Central Participants

Of the six focal students I followed, three participated more centrally in their classroom CoP than the others. The three who became central participants (according to Lave and Wenger's terminology) were Thomas, Pedro, and Anna. Each of these students participated actively (within either whole class or small group configurations), carried out written assignments in class appropriately (thus participating in classroom literacy practices), and succeeded on in-class and standardized tests. To illustrate my findings, I will show how each of these 'central' participants interacted within their classroom CoP, using higher order language and a range of academic language functions. It should be noted, too, that Thomas, Pedro, and Anna also succeeded in traditional measures of school success (in-class work, class tests, standardized tests).

To illustrate how these three focal students used a diverse range of discourse, as well as a rather high degree of complexity of discourse within their communities, I present a characteristic interchange between the focal student and another classroom community participant—either the teacher or another peer.

Thomas

In Example 5.1, Mr. Martinez draws attention to Thomas' strategy for identifying which fraction is larger or smaller when comparing fractions. Within this example, Thomas demonstrates higher order thinking in his response to the teacher.

In Example 5.1, Thomas uses higher order thinking throughout his discursive interaction with Mr. Martinez. Although Mr. Martinez has elicited language from the class as a whole, Thomas jumps in, as he often does, to respond to the teacher's questioning. Line 2 shows Thomas analyzing the teacher's question and responding. In line 3, Mr. Martinez again addresses the class as a whole, but Thomas has garnered the attention of his teacher.

EXAMPLE 5.1
THOMAS COMPARES FRACTIONS, JANUARY 28

M = Mr. Martinez
T = Thomas

1. M: *Qué está usando Thomas para figurar la respuesta aquí?* [What (strategy) is Thomas using to figure out the answer here?]
2. T: The higher number!
3. M: Okay what strategy is he using here?
4. T: That the seven is higher than the three.
5. M: The what?
6. T: That the seven is higher than the three.
7. M: Okay, but why are you saying that?
8. T: Because the twelve is the same so, it's xx.
9. M: Good, the denominator, with the denominator the same, all you have to do <is compare the numerators> [>].
10. T: <Is look at the top> [<].

In lines 4 through 8, Thomas is engaged with the teacher and provides responses which reflect use of higher order thinking, giving reason or rationalizing, in particular (lines 4, 6, 8). Mr. Martinez recaps the main point of the discursive segment in line 9. In the last line, line 10, Thomas speaks at the same time as the teacher does and adds a final higher order evaluation of his strategy to which Mr. Martinez has drawn the classes' attention.

This characteristic example of Thomas' discourse illustrates how he uses higher order thinking skills in his language when interacting with the teacher, his primary interlocutor in math class. It further shows him as a central member of the classroom community of practice. Thomas is participatory and demonstrates ownership of mathematical discourse through higher order language and interaction in class. He is a leader whose participation is central in the classroom community.

Pedro and Anna

Like Thomas, Pedro was a leader in his class. He engaged in whole class interactions, as well as in more intimate, small group configurations. Although Anna was often shy in whole class settings, she was usually comfortable working within small groups, especially when she could work with Pedro. Anna and Pedro's diverse use of language, both simple and complex, can be seen in the following example. On February 23, Pedro and Anna are working with their group members, Sarina and Emily. They have split into two groups of two, and are trying to determine the total number M & Ms they had in their piles so they can write ratios and fractions.

EXAMPLE 5.2
PEDRO AND ANNA ANALYZING PROBABILITY OF M & Ms, FEBRUARY 23

S = Sarina
P = Pedro
A = Anna

1. S: We have twenty-four.
2. P: Eh, you have different?
3. E: Yeah, twenty-five.
4. A: No, we have one two three four five six seven eight nine ten eleven.
5. A: twelve thirteen, fourteen, you have twenty-five I just counted!
6. P: I got twenty-five.
7. E: Yeah, you got twenty-five, divide twenty-five into one.
8. P: But like this. (he shows her (with M &Ms) how he got the answer)
9. P: It's ten, twenty-five, thirty, it goes in <xx times> [>].
10. A: <I have twenty-five> [<], xxx it's twenty-five, look!
11. P: It's twenty-four?
12. A: Five, look, one, two, three four five six seven,< eight> [>], nine, ten.
13. P: <I know> [<] but Miss Koch said it would be [:a] fraction.

In Example 5.2 Pedro and Anna participate actively in their small group and use a diverse range of language functions to negotiate the problem of analyzing probability.

Sarina declares that she and Pedro have twenty-five. In line 2, Pedro uses questioning to check what Emily and Anna have come up with. Emily concedes with Pedro that she has a different answer than Sarina does (line 2). Anna then engages in the discourse, using reasoning or rationalization to explain her calculation (lines 4, 5). A few seconds later, Pedro has refined his answer and uses lower order language to describe his solution (line 6). Emily concurs with him (line 7). In line 8 Pedro defends his calculation and aims to explain how he got it by providing reason. He then describes with numbers how he arrived at the solution he did (line 9). Anna returns to the discourse with a description of her solution (line 10). Pedro questions her solution in line 11. Anna maintained she had twenty-five, but Pedro understood she had twenty-four, and poses this as a question. Although his question is not phrased in a standard question format in English, he has clearly communicated to Anna his intent. In line 12, Anna again defends her answer. She uses the higher order skill of reasoning by showing her group how she arrived at the solution. In line 13 Pedro uses reasoning, citing that the teacher said the answer they were to provide must be in the form of a fraction.

In Example 5.2 Pedro has used both lower order language of describing, as well as higher order language of questioning and reasoning. His discourse shows that he is an independent learner who monitors his understanding by checking with peers when concepts are unclear to him. He is also able to defend his responses by reasoning and rationalizing. Pedro's use of discourse in math shows that he is a productive and efficient learner who can use a diverse range of discourse features in his practice. He is a learner who experiments with situated academic language, but who also has taken ownership of many discourse practices that can help him learn mathematics.

Anna also experiments with mathematical discourse. She uses the higher order function of reasoning to explain and defend her answer to her peers, but when she lacks the language to verbalize her rationale, she presents her calculations to illustrate her intent. She uses lower order description when negotiating with peers, and uses higher order language to reason in group situations. Like Pedro, Anna is also a productive and diverse user of mathematical language.

Thomas, Pedro, and Anna used critical thinking in their mathematical discourse. They were also creative and diverse in their use of language. These three practiced both higher and lower order thinking discursively in math class and participated in ways that reflected confidence and ownership of the language they used to communicate and interact about math. With respect to Halliday's (1978) language functions, these three ELLs used a variety of important functions necessary for school. Thomas often utilized regulatory, informative, and heuristic functions, while Anna and Pedro used a good deal of informative, interactional, heuristic, and also imaginative functions in their classroom discourse—as a result of the common group work and frames of play in their class. Although Pedro and Anna were still classified as ELLs, and occasionally struggled with the challenges of talking in and about math, their performances of participation in class showed they could use math language in a variety of ways. They used it receptively, and experimented with it when attempting to take ownership of it. Thomas, Pedro, and Anna were successful in math class partly because of the ways in which they utilized mathematical discourse.

'Legitimate' Peripheral Participants

Jennifer, Nestor, and Benjamin, did not exhibit the diversity, nor the use of higher order thinking in discourse that the central participants, Thomas, Pedro, and Anna did.

Next, I will illustrate this claim by presenting a characteristic sample of 'legitimate' peripheral students' language.

Jennifer

Jennifer was something of an anomaly, as her participation could be active and engaged, but she generally did not do well on traditional measures of school success. On some days, and in particular, on the days in which group work was utilized in class, she participated actively and successfully. Her use of discourse practices ran the gamut from lower to higher order, and she could demonstrate proficiency in using some higher order discourse practices. On February 3, she is working with her bilingual girlfriend, Kristina, on tripling recipes of brownies.

EXAMPLE 5.3
JENNIFER GIVES DIRECTIONS ON BROWNIE TASK, FEBRUARY 3

J = Jennifer
K = Kristina

1. K: Okay, um, what do we do now?
2. J: Now we're gonna times it by +/.
3. K: By two.
   ——(a few seconds later)
4. J: I'll do, xx there's eleven.
5. J: So we'll do, I'll do two and you'll do three.
6. J: I'm thinking about money, one, two, three for me.
7. J: Yeah look, from one eight down, you do, one fourth, from one cup +/.

In this example, the discourse features Jennifer uses most are describing and giving simple directions to Kristina. Jennifer answers her partner's question with a description of the process in line 2. A few seconds later, Jennifer begins to parcel out each partner's task. In line 4, she describes the amount of problems they must do, and in line 5 she gives Kristina instructions regarding her task. Jennifer continues to describe her thought process about how the assignment will be carried out, by counting the problems out (line 6). In line 7, she uses mathematical discourse to give directions to her partner about which parts of the assignment Kristina will do.

This example of Jennifer's interactive discourse shows how she uses discourse features to participate in group negotiations. On this day, many of the discourse features she used were lower order, but her mathematical discourse practices included a range of features she could access in order to communicate. When she was presented with opportunities in which to work with bilingual girlfriends, Jennifer played a central role in group negotiations. On these occasions, her mathematical discourse could be rich. Individual tasks (which were usually assessments on which she was graded) were much more challenging for her.

Nestor

Nestor did not benefit from group work as Jennifer did. When he worked in groups, he attempted to utilize mathematical discourse practices, but his attempts to engage others often fell on deaf ears. In this example, Nestor tries to understand the task he and his partner, Jon, must do.

EXAMPLE 5.4
NESTOR ASKS FOR HELP ON GRID, FEBRUARY 4

L = Ms. Loesely
N = Nestor

1. L: Underneath your grid, write what you shaded.
2. N: You count it?
3. N: Of a hundred, ten, twenty, thirty, forty, fifty, sixty, seventy, eighty.
   (counting very quietly to himself)
4. L: And you can write the decimal of twenty-two hundreds, can you write the decimal, twenty-two out of a hundred, talk it over in your group, see if you can figure out how to write the decimal.
5. N: How do you do it? (to Jon, his partner)
   (Nestor does not receive a response)

After Ms. Loesely has given the instructions (line 1), Nestor uses confirmation to understand what he needs to do (line 2), but he does not receive a response. In line 3, Nestor begins to attempt to work out the problem by himself, by counting aloud, a feature of lower order discourse practice. The complete utterance in line 4 shows Ms. Loesely modeling mathematical discourse and also encouraging the students to utilize math discourse when figuring out how to write the problem. However, we see in line 5 that Nestor is still unclear about the task, and although he overtly elicits help, he is not recognized.

Example 5.4 reveals Nestor's characteristic discourse. In class, Nestor was quiet, resistant to working in groups, and generally quite invisible. Although he utilized some useful features of discourse to attempt to engage others in interaction, he did not receive the recognition he needed.

In general, Nestor did not use a great deal of higher order language, nor did he engage in a variety of language functions. Much of his linguistic interactions were at the level of passive and receptive learning. He did not have the opportunities for participation to develop his mathematical discourse practice to an advanced level. Thus, Nestor did not move from 'legitimate' peripheral participation to central participation in the CoP in his classroom.

Benjamin

Like Nestor, Benjamin lacked opportunities to practice his mathematical discourse. However, Benjamin did not have the same challenges that Nestor faced. Whereas Nestor was not effective at engaging his discursive partners, Benjamin did not have the opportunities to work one on one with others since Mr. Martinez did not usually incorporate cooperative learning in his class. In the following example from January 28, Benjamin was provided with an opportunity to work with Thomas.

EXAMPLE 5.5
BENJAMIN GETS HELP COMPARING FRACTIONS, JANUARY 28

B = Benjamin
T = Thomas

1. T: Two-fourths is greater than two-fifths, right?
2. B: Okay, then, this one is?
3. T: Okay, reduce it!
4. T: Reduce this by three!
5. B: By three, okay.
6. T: So it's three, um, three +/.
7. B: Three six?
8. T: Three-sixths!

As Example 5.5 begins, Thomas confirms that Benjamin understands a simple fraction comparison. Benjamin acknowledges he understands, and then uses questioning to check how to carry out the next problem (line 2).

Lines 3 and 4 show Thomas giving stern directions to Benjamin. In line 5, Benjamin acknowledges Thomas' directives by revoicing Thomas' words. In line 6, Thomas is calculating aloud, and Benjamin finishes his utterance, by calculating, a feature of discourse classified here as higher order. Thomas corrects Benjamin's pronunciation by revoicing his contribution.

Although Benjamin was often quiet, he used some discourse practices in math that helped him think critically and learn mathematical concepts. However, the majority of his interactions in class were lower order, and much of his learning was passive or receptive. He simply did not have enough opportunities to engage in classroom interaction that he needed in order to shift from a peripheral participant to a central participant.

Conclusion: Student Participation

In comparing the two groups of focal students as 'successful' and 'unsuccessful' or central and peripheral, in terms of utterances spoken in class, the central group generally spoke more often than the peripheral group. In terms of complexity of discourse use, the central participants employed more higher order thinking discourse practices than the peripheral group, and they used a variety of language functions in their discourse, especially those which are exclusively school-related, such as the heuristic.

Having a picture of how these ELLs practice discourse can show how their discourse develops, but it is critically important to understand why some of these students became central and ultimately 'successful' in math, while others remained peripheral and ultimately 'unsuccessful'. I argue that three factors were crucial to centrality and 'success.' First, opportunities for participation were critical to the development of mathematical discourse. Second, math skill played an important role in whether a student began to take ownership of math discourse. Third, students needed specific training and guidance in using and developing their discourse practices.

Nestor and Benjamin lacked the opportunities that Thomas, Anna, and Pedro had to practice mathematical discourse. Certainly both Ms. Loesely and Mr. Martinez encouraged and expected Nestor and Benjamin to participate, but due to resistance on the part of the student, or lack of teacher-instituted opportunities for interaction, these boys did not utilize math discourse to the extent they could have. Jennifer had opportunities, but she lacked math skill and training. They all needed more interaction, as well as more opportunities to develop and use heuristic, informative, and higher order language skills.

Whereas Thomas, Pedro, and Anna had a strong background in math, Jennifer, Nestor and Benjamin did not have a firm grasp of basic skills in math, and the ability to apply them to the classroom setting. Their lack of skill contributed to their inability to utilize diverse discourse practices in class.

Finally, Jennifer, Nestor, and Benjamin would have benefitted from explicit training in mathematical discourse. Certainly, Ms. Loesely often emphasized features of mathematical language, but Jennifer and Nestor may not have been engaged. In Mr. Martinez' class, Benjamin did not often elicit language or emphasize features of mathematical discourse. These students needed more focus on math discourse practice.

Participation Performances

In this final section of this chapter, I describe how the focal ELLs in my study demonstrated the process of development of mathematical discourse learning. I use analyses of student and teacher discourse in this chapter to address my primary question of how language learners develop math discourse in reform-oriented mathematics classes.

Over the course of the five months that I regularly spent in the classroom, I noticed that some of the bilinguals and ELLs followed a similar pattern of new mathematical discourse development. The students exhibited different features within the learning process, starting with a passive or receptive type of learning, then progressing to a more advanced experimentation, before taking full ownership of the mathematical language.

These participation performances, as I am calling them, are similar to stages, but do not manifest themselves in a strictly sequential fashion as a traditional stage would. The sixth grade ELLs passed through some of the various kinds of participation performances in a somewhat recursive, rather than linear manner. As a result, I refrain from using the term *stage*, which evokes linearity and chronology. This said, I have discovered that the three central and 'successful' focal students, Thomas, Pedro, and Anna, participated more commonly in performances of advanced experimentation and ownership, whereas the peripheral and 'less successful' focal students, Jennifer, Nestor, and Benjamin participated more in passive or receptive and experimentation performances throughout the semester. Table 5.2 summarizes the three participation performances that I have developed on the basis of my analysis of focal student language.

Passive/Receptive

The students who were at a lower stage of language proficiency in English, Anna, Benjamin, and Nestor (based on standardized test scores, all three of these students were classified as beginner or low intermediate ELLs) exhibited the most distinct features of the first type of participation performance. These three focal students were keenly aware that English

TABLE 5.2 The Process of Mathematical Discourse Development

What types of participation performances characterize the process of mathematical discourse development by ELLs?

Types of participation performances
Passive/Receptive learning
Appear to understand directions and content learned previously, but hesitant to use the newly introduced structures and vocabulary. Code-switching occurs in interactions. Utilize learning strategies such as repeating new vocabulary. Use of short, uncomplex Reponses, ending with a rising, question-like intonation
Experimentation
"denom, whatever!", and also confuse terms like multiple and multiply. Better able to use polysemous vocabulary and homophones. They become accustomed to using common words like table, bar, and line that have context specific meanings. They negotiate understanding and production of target forms with others.
Ownership of the discourse and content
Learners are comfortable with use of newly developed discourse forms. They integrate the discourse into mathematical discussions in small groups and whole class formats. Discourse matches closely the target form of teacher and texts.

was the language of school, and in math as well. I observed the passive/receptive learning performances most frequently in the early part of my data collection (beginning of spring term).

While their shy demeanor undoubtedly played a role in their hesitance to utilize new math discourse, two of the three limited language learners, Nestor and Benjamin, displayed reluctance to interact discursively with others, especially when they were expected to be "talking math." When called on by the teacher, all three limited English speakers were usually extremely hesitant to respond in front of the class. When working with partners or in small groups, Nestor and Benjamin participated somewhat, but their participation was characterized by revoicing of their interlocutors and repeating of terms and concepts, short utterances, and responses ending with rising intonation, like a question. Although she admitted she was shy, Anna was participatory in her small group, especially when was with Pedro.

Nestor's hesitance to speak in class can be partly attributed to his lack of skill and confidence in his math ability. In Example 5.6 Nestor and his partner are completing ACE (Applications-Connections-Extensions) questions from Investigation 3: Cooking with fractions from *Bits and Pieces I.* The students have been given the tasks of illustrating a fraction in a grid sheet "by drawing a square, subdividing it into equal size regions, and shading the fractional part indicated" (Lappan et al., 2002a, p.34). The first problem Nestor is working on is 7/20. His answer looked something like what is shown in Figure 5.2.

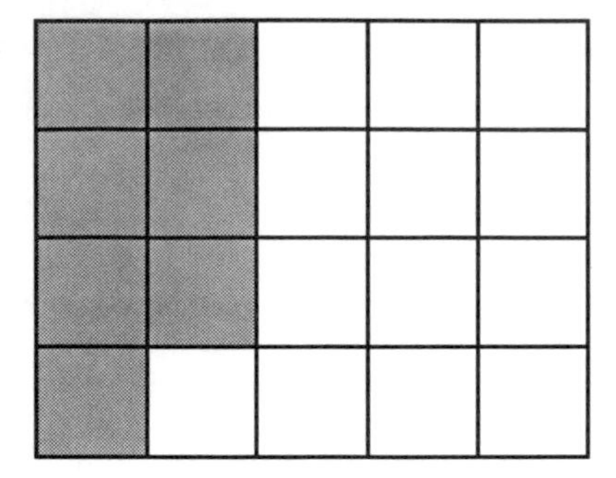

Figure 5.2 Nestor's representation of 7/20ths.

He has successfully solved this problem on paper. However, when I asked him what he had done, he was unsure about describing his process. Example 5.6 shows our interaction.

EXAMPLE 5.6
NESTOR AS A RECEPTIVE LEARNER, FEBRUARY 2

N = Nestor
H = Holly

1. H: What are you doing?
2. N: Um, this (points to grid sheet and book).
3. H: Can you tell me what you are doing?
4. N: Um, on the shade, um, decimal, and + . . .
5. H: How come you shade seven?
6. N: Um, there's seven and twenty.
7. H: What's that called?
8. N: Um.
9. N: I don't know.

While he got this particular problem correct, Nestor's language did not reflect his ability to explain his solution. His performance later in the class also showed that he did not completely understand how to apply his knowledge from this problem to the others. The next problem tells students to carry out the same process for 3/15ths. But Nestor shaded the entire square of fifteen pieces. His partner, Jon, explains to him that he got it wrong, that what he had shaded "would be like fifteen-fifteenths" and that Nestor needed to shade only three pieces.

Benjamin was not actually called upon often by his teacher, Mr. Martinez, but when he was, Benjamin usually responded with his common refrain:

"I dunno." Like Nestor, Benjamin was also lacking skill and confidence in math. Example 5.7 from January 28 shows Benjamin working with Thomas. Thomas is helping his friend understand how to compare fractions.

EXAMPLE 5.7
BENJAMIN GETS HELP COMPARING FRACTIONS, JANUARY 28

B = Benjamin
T = Thomas

1. B: How do you do it?
2. T: You reduce it!
3. B: ### (silence).
4. T: That means, divide it!
5. B: Yeah?
6. T: No, no, reduce it, I forgot how to say, hold on.
7. B: I dunno.

Benjamin's discourse in Example 5.7 is characterized by short utterances, a silent pause, questions, a plea for help and his common refrain, 'I dunno'. His negotiation with Thomas showed that he was engaged in the interaction, as he was following along, but he was still quite reluctant to use complex mathematical discourse.

When presented with new concepts and terms, or with mathematical notions out of context, Anna would practice them under her breath, as self-talk. She appeared to understand much more than she could produce, as evidenced by her engagement and successful participation in activities and assignments. On February 23, Anna and her class had been working on probability for over a month, and were constructing a pie graph to represent colors of M & Ms she and her partners discovered in one bag. When Ms. Koch explains that the students need to know "how many degrees each (angle of the spinner) would be" Anna first questions "*degrees?*" and then goes on to repeat the term two more times. Since the class had already studied the CMP unit on shapes, she knew the meaning of the term *degrees*. Her reaction to Ms. Koch's (re)introduction of the term suggested that she was using the strategy of repeating to gain a better understanding of the term. She may have been reviewing the meaning or sorting a homonym in her mind, noticing that the term *degrees* is used to measure angles, but also to measure temperature.

Of the three limited focal ELLs, Anna was the most participatory and interactive in small group work. Anna often code-switched between English and Spanish when getting familiar with new discourse and when working with group members with whom she knew could understand Spanish. However, her code switching appeared to occur during passive/receptive performances, as well as in the middle level of the three participation performances—the one I have named experimentation.

For much of the semester, Nestor and Benjamin's discourse development was in the passive/receptive participation performances. Although they displayed some more advanced performances of participation late in the year, the majority of their discourse practices remained at the less advanced phases of development.

Experimentation

The experimentation performance is where many of the focal students demonstrated their discourse practices. The students who were more proficient in both English and math, such as Thomas and Pedro produced some statements indicating clear evidence of learning, but at the same time, were still 'getting used to' the math discourse. Less proficient English language learners also exhibited characteristics of this experimentation with mathematical discourse. I saw examples of this performance throughout the period of data collection. In an example excerpted from February 13 (seen in Example 3.10) Pedro is trying to answer his teacher's question, but he becomes briefly tripped up on a rather challenging, yet commonly used five syllable word: *denominator*. He asks Ms. Koch: "The *denominer*, the *denominator* is still one hundred and twenty-five, right?" Pedro was familiar with the word *denominator*, as he had used it in other instances in class. Not to mention the fact that his teacher modeled the term numerous times within the probability unit they were working on, as well as in the introductory unit on fractions. It appears here, then, that Pedro is developing ownership of this important fraction term, but is still working to integrate it into his vocabulary and/or phonology.

Mr. Martinez' class was introduced to basic fraction vocabulary (such as *numerator* and *denominator*) very early in the school year (and likely also in elementary school), but they were still challenged by these terms (with regard to meaning or pronunciation) throughout the semester. As late as March 4, Thomas and Benjamin had not learned how to appropriately use these important terms. Thomas was helping his friend to find the LCD (least common denominator) and asks: "You write that one up there with the five, so that's um, your *nominator.*" To which Benjamin agrees, "*nominator*, yeah." Thomas then confirms with "Okay, put it, the *nominator!*" We see here, that while Thomas is the best student in the class, he is also a language learner, still experimenting with mathematical discourse usage. His identity as a "good student" often protected him from being challenged by other students and by the teacher.

When experimenting with target forms of mathematical discourse in English, focal students worked in concert with classmates to negotiate language and understand concepts. I will illustrate in the examples that follow how some of the focal students negotiated their learning of content and discourse by experimentation and reliance on group or peer interaction.

Jennifer was classified as no longer needing language support in school. Thus, while her language proficiency in English was advanced, her mathematical knowledge was not at the same level. In the next example, I provide a typical example of Jennifer experimenting with mathematical discourse. Throughout the semester, most of Jennifer's participation remained at the level of experimentation.

On February 16, Jennifer and one of her bilingual friends, Patricia, were working as partners on an activity in which they needed to identify characteristics of cats and ultimately derive percentages according to the fractions they came up with. Ms. Loesely was gone that day, so I was left with the substitute teacher. Since I was a familiar face on campus and in the classroom, I was permitted to stay and talk with students about their math learning, even when the teacher was not there. The girls had come across a division problem with an answer that had repeating elements (i.e., 3.333). When pressed by me, Jennifer remembered that it was necessary to add a line over the repeating numbers, but had some difficulty verbalizing what the symbol meant. The following except illustrates Jennifer's ability to recognize the meaning of the symbol, but also her uncertainty in defining it.

EXAMPLE 5.8
JENNIFER AND PATRICIA AND REPEATING NUMBERS, FEBRUARY 16

J = Jennifer
P = Patricia
H = Holly

1. J: Another three and then put a line over it cuz it just keeps going.
2. P: And going and going and going!
3. H: What is that called?
4. J: It repeats itself.
5. H: Repeats, yeah, okay, okay.

In this example, we can see how the students work together to negotiate the linguistic referent for the repeating symbol. At the same time, however, it is apparent that the girls have not yet gained ownership of the term. About five minutes later when Patricia is checking over her work, she mentions: "Ah, I forgot to put that little stick!" When I elicited further clarification, Jennifer came to her rescue and revoiced and defined Patricia's intent: "the little stick, it's the little line on top."

What is most interesting about this example is not that Jennifer and Patricia encountered difficulty with the label (as there does not seem to be a commonly used word for the repeating bar—according to Dr. Math (Peterson) at the mathforum.org), but that they struggle with describing its purpose. When working together, or with a more experienced tutor or peer, however, the girls manage to come up with appropriate definitions and labels for the important terms and concepts.

In a similar instance, during the second interview, conducted in a group format on May 11, Thomas and his friend Armando are explaining how to multiply percentages and come across a repeating bar. As they explicate the process, they piggyback off each other and produce a descriptive explanation for the concept. Thomas has reached the end of the problem and begins to describe a repeating number: "It's continuous, it never ends." he says. Then Armando supports his friend's claim and provides further information: "You put a dash." To which Thomas adds: "That's the endless part!" Their feeding off each other within the joint interaction worked to produce a coherent explanation of repeating numbers. Although Armando's use of the term *dash* instead of more commonly used words for this concept such as *line* or *bar* is less than conventional, he nonetheless conveys the concept and understanding of how to represent it in written form.

Ownership

In what I am calling the ownership performance, ELLs have a firm, confident grasp of the new terms and structures of fractions, and demonstrate this not only by teaching, but also, correcting, their peers on discursive issues. Throughout the semester, the students who most often exhibited performances of ownership were the central, 'successful' students, Thomas, Pedro, and Anna. Pedro and Anna demonstrated ownership of a range of discourse practices they used to determine if a particular problem they had completed was correct. As the example begins, Anna has just asked Pedro what he did wrong.

In Example 5.9, Anna and Pedro demonstrate they have ownership over the discourse practices that help them understand how they solved the probability problem. They use both higher and lower order discourse practices to negotiate a solution. In line 1 Pedro evaluates the problem and determines there may be a problem with the solution. Anna then begins to rationalize what they had done (line 2). In line 4, Anna explains their calculation and then, using a higher order function, analyzes what she thinks the answer is. Pedro does the same, and begins to display his reasoning, as the example closes. In the end, Ms. Koch helps the two discover that they have

EXAMPLE 5.9
ANNA AND PEDRO DEMONSTRATE OWNERSHIP OF DISCOURSE PRACTICES, FEBRUARY 23

A = Anna
P = Pedro
K = Ms. Koch

1. P: I think we did it wrong because it came out to eighty-two percent.
2. A: Yeah, miss, cuz we add +/.
   (teacher comes over)
3. K: Let's see what you got.
4. A: We add[ed] thirty-two plus sixteen plus two plus twenty-eight.
5. A: Miss, I think it's twenty-five over twenty-five.
6. P: That's what I put, I put um+/.

come up with the correct answer. This example nevertheless demonstrates that Anna and Pedro have developed ownership over discourse practices that help them understand how to solve and correct problems.

In the following example from early in the spring term, Thomas is helping Benjamin to understand comparing fractions. Benjamin wants to begin analyzing the fractions, but hasn't simplified his fractions yet. Thomas comes over and assists him. Armando is sitting next to Benjamin, so he jumps into the conversation.

EXAMPLE 5.10
THOMAS, BENJAMIN, AND ARMANDO REDUCING FRACTIONS, JANUARY 28

B = Benjamin
T = Thomas
A = Armando

1. B: How do you do it?
2. T: You reduce it.
3. T: That means, divide it.
4. B: Yeah?
5. T: No, no, reduce it, I forgot how to say, hold on.
6. B: I dunno.
7. A: What are you talking about?

8. T: Oh, now I remember.
9. A: What's on your paper, Thomas?
10. T: Look, reduce it by three, divide by three, it's too long.

In this sequence, Thomas indicates that he knows how to simplify the fraction—as evidenced by his first response to Benjamin. He is even able to describe how to reduce a number—by dividing it. Thomas has developed ownership of the meaning and usage of *to reduce* in math. At this early stage in the semester, however he may still be experimenting with other forms used to express the same operation. Within the fraction unit, Mr. Martinez uses terms and expressions such as *lowest terms, simplify,* and even *GCF, Greatest Common Factor,* and *LCD* or *Least Common Denominator* to refer to reducing or reduced numbers.

Several months later, on March 4, Thomas is multiplying fractions and needs to be sure the fractions are in simplified form. For the majority of the class, Thomas is working alone but thinking aloud. On several occasions, Benjamin or Armando call on him for help. Within one 48-minute lesson, Thomas uses a form of *to reduce* (including *reduce, reduced,* and *reducible*) 18 times when working through the fraction problems. While he does not use *simplify* or *lowest terms,* he indicates his comprehension of these terms when he is confronted with them by either Benjamin or the teacher (respectively).

On May 13, Benjamin also demonstrates proficiency with the term *reduce.* Not only was he able to use the term and the concept with confidence, overall, he interacted and participated a good deal more in math class than he had done before. He demonstrated ownership of certain syntactic constructions such as *x goes into x,* and *what's x times x?* He was also more confident in eliciting assistance from other more capable peers in class. Several times within the lesson, Benjamin called on Armando, Thomas, and the teacher for help. However, he had not taken ownership of the concepts. When Thomas came to help Benjamin reduce a problem, Thomas realized that Benjamin had done that particular problem, and others previous, all wrong.

Benjamin's active participation is likely a result of being more comfortable speaking in class as well as becoming accustomed to working with his classmates within Mr. Martinez' classroom dynamic. Quantity of interaction and engagement in math class also likely played an important role. At this point in the semester, all sixth-grade students at Hot Springs had just completed a several week period in which they had two math classes per day instead of an elective. "Double math" as it was called was intended to prepare students for the state standardized test. Although Benjamin's enthusiasm and participation in daily lessons appeared to increase toward the end of the semester, he was still lacking critical mathematical skills and concepts, as evidenced by his questions of "What is three times nine?" and "Can you reduce 18 over 20?" It was clear that, in addition to his lack of academic language knowledge, Benjamin also needed practice and much closer guidance to be able to become a proficient speaker and user of mathematics.

In much the same way that Benjamin participated more fully and began to take ownership of mathematical discourse, Nestor also 'woke up' toward the end of the semester. Example 3.9 from chapter three showed how Nestor was able to describe the conversion of fractions to decimals, as well as how to apply it to get correct answers. In the case of Nestor and Benjamin, more time, exposure, and participation in mathematical discourse was necessary for them to develop ownership of many features of discourse.

When working in groups, with experienced peers, or with the scaffolding of the teacher, the students often produced the appropriate mathematical discursive features. In some cases, however, the group interaction would have benefitted from closer monitoring and scaffolding by the teacher (March 4, Thomas and Benjamin, *nominator*). In general, when students had the opportunity to interact, negotiate, and monitor each other in math, they had more, richer experiences in which to utilize the mathematical discourse and participate in the different kinds of performances that they did, and ultimately more opportunities to attain ownership of the discourse.

CONCLUSION

This chapter has explored the question of what mathematical discourse for sixth graders is, as well as how language learners develop that discourse in CMP classes. Some focal students participated in the center of the community of practice which emerged in the classroom and ultimately achieved success in traditional measures of school and in group participation, while others did not.

A number of factors play into whether a language learner is successful at developing and producing mathematical discourse or not. In general, I found that many of the students I followed were creative and intelligent thinkers when using school and math language. However, students who became central participants were those who had opportunities for participation, a good grounding in basic mathematical skill, and training and explicit attention to discourse. Although some students who remained peripheral were given opportunities for participation and attention to discourse features, they were resistant to these opportunities.

When learning, students gained competence in, or 'ownership' of discursive features of mathematical language by experimenting with discourse features, vocabulary, syntax, and expressions. Focal students who remained peripheral did not leave the participation performances of passive/receptive or experimentation. Students who became central participants were more successful at developing ownership of mathematical discourse. Teachers focused on mathematical discourse to help learners develop competence in math and math discourse, but clearly individual factors such as math skills, appropriate guidance, and opportunity for rich, academic participation in class proved to be an important part of successful learning in the sixth grade classes I followed.

NOTES

1. It should be noted that b cannot be equal to zero. The citation above comes directly from *Bits and Pieces II*, 2002b version. I thank Dr. Betty Travis of the University of Texas at San Antonio for this observation.
2. I acknowledge the updated version by Anderson and Krathwohl (2001), but have chosen to reference the original for simplicity.

CHAPTER 6

CONCLUSIONS AND IMPLICATIONS

In this chapter, I address important issues raised in my study with regard to teaching and learning math with Latina/o ELLs in an academic environment that stresses participation, problem solving, and discursive interaction. I revisit my original research questions presented at the beginning of the book, and draw implications based on the conclusions I present. As a reference, I list them again here.

- How do language learners develop math discourse and gain math content knowledge in a reform-oriented classroom?
- What is the effect of participation in the classroom community on the successful use, development of, and socialization to math discourse and content knowledge?
- How do individual variables such as language proficiency, gender, knowledge of math and the 'ways of school', and academic identity affect participation in math? and
- What features are present in reform oriented math curricula, or those classes which utilize this curricula that facilitate learning of mathematical discourse for ELLs?

In this concluding chapter, I address how individual variables affect participation in math. Next, I address the effectiveness of reform math for

English Language Learners and Math, pages 121–135

urban, Latina/o middle school ELLs and bilinguals and discuss how CMP is beneficial for these students, citing the features of the curriculum and its presentation that help them to engage and participate in math. I then discuss how participation in a CMP classroom can (and cannot) explain learning of mathematical discourse in reform-oriented classrooms. Finally, I conclude with implications and limitations of this study, and directions for further research.

INDIVIDUAL VARIABLES AFFECT PARTICIPATION

To understand how middle school Latina/o ELLs developed mathematical discourse in reform-oriented math classrooms, I considered how gender, ethnicity, knowledge of math and school, language proficiency, and teacher practices as related to reform math learning. In Chapters 2–5, I discussed the individual focal students in terms of individual variables. These chapters suggest ways in which these factors affect participation.

Gender

When speaking with students in one-on-one interviews, one of the questions I asked was whether they believed that boys were better in math than girls (or vice versa). Interestingly, every one of the respondents claimed that gender did not play a role in one's ability in math class. Jennifer indicated that being a girl helped her "get away with a lot of stuff" in math class, but she felt, overall, that girls and boys had similar capabilities. Anna felt that gender did not affect one's abilities in math. In an interview she explained at length: "It doesn't matter because there's boy [= who are] intelligents and there's girl [= who are] intelligents... there are some girls that are not intelligent and boys that are not intelligent." On the same day, she revealed her pragmatic attitude to gender and achievement when she shared her mother's wisdom: "My mom told me, *una mujer puede hacer lo mismo que un hombre* [a woman can do the same as a man], but he [= she] have to try more than the man." All of the boys I spoke with believed that girls were equally capable as boys. Only Pedro suggested a difference in genders. "The girls talk more" he said. He explained that the increase in talking wasn't necessarily math talk, but social talk.

Pedro's argument had a grain of truth to it. Where gender did affect participation was within group work. According to Pedro, the girls were much more talkative than boys in group situations in class, as well as out of class. For Jennifer, talking with girls facilitated her understanding of fractions and use of mathematical discourse. Jennifer indicated that she felt more comfortable working with her girlfriends on math activities. She had a large group of bilingual friends (primarily girls) that supported her in class, and their camaraderie helped her complete assignments in class. Besides her interactions with Pedro, Anna worked exclusively with girls in her math class. Thomas also appeared to work best with boys. He helped them more often than girls, but could also be seen helping girls when requested. Nestor and Benjamin also usually chose to work with the same gender, when they had the opportunity to work in small groups. However, Pedro was amenable to working with just about anyone, as long as they could keep up with him.

None of the teachers believed that at the sixth grade level, one gender was more capable than the other of doing math. Within the interactive nature of CMP lessons, students are expected to engage in a variety of grouping configurations, including whole class, small group, and pair work. Each of the six focal students—both boys and girls—highlighted in this study was able to comply with this expectation.

Overall, with respect to capability, there was no difference between the genders. Students did, however, choose patterns of participation that were gender marked. Girls wanted to work with girls and boys wanted to work with boys. To determine whether Pedro's comment that girls talked more than boys would need to be answered within a study of a different type.

I acknowledge that gender may in fact pose a problem for some of the focal students later in their mathematical careers, but according to research on gender and education, a clear, discernable gender difference is not likely to arise until after the age of eleven or twelve (Gilligan, 1982; Sadker & Sadker, 1995). None of the students I worked with, with the exception of Jennifer, had yet reached this developmental milestone during the time I spent with them in math class.

Latina/o Ethnicity

Teacher ethnicity appeared to have affected student participation in Mr. Martinez' class. Since he shared the same ethnic background and language with the majority of the students in class, he could and did incorporate references to their shared ethnicity and Latina/o identity in the lessons. By so doing, he made connections with the students and motivated them to engage in the discourse.

Although she did not speak Spanish, or share ethnicity with most of her students, Ms. Loesely occasionally used Spanish to connect with ELLs and bilinguals. She could not give instructions, model mathematical discourse, or explain concepts with the students' L1, but when she used short phrases in Spanish, students responded.

Where the teacher or curriculum failed to address the ethnicity of Springvale's students, students filled in the gaps. In Ms. Koch's class on March 23, Pedro, Anna, and their group are choosing what to order at a restaurant. When Pedro reads 'Shrimp cocktail' he translates it as *Cocktail de Camarón* and requests that it be served with *pico de gallo*, a condiment comprising tomatoes, onions, and peppers. His partners think this is quite funny. Nevertheless, the white teachers did not have the same advantage that Mr. Martinez did—in using it to promote student participation and interaction—since they shared neither ethnicity nor language with many of the students in their classes. Ethnicity affected participation most in Martinez' class, where more students could connect with their teacher's ethnic identity.

Knowledge of Math and of School

As discussed in chapter five, mathematical skill and knowledge of school played an important role in the students' success in math. The two most successful focal students, Thomas and Pedro, had both been educated in the U.S. all of their lives and had made good grades throughout elementary school. They both had skill and knowledge of how math class worked and how they needed to interact within it. Thomas and Pedro were recognized as good students by their teacher and by other students. They often chose to flaunt their skill and ability by participating actively and vociferously in class. In addition, they, along with all of their sixth grade classmates who had attended elementary school at Springvale schools, were familiar with reform-oriented mathematics. This familiarity with the presumptions of reform math helped Thomas and Pedro to participate and engage in class. As a result, they had many opportunities in which to practice the mathematical discourse they were learning.

Although Anna was somewhat new to the American educational system, she was knowledgeable in math and was proficient in basic mathematical concepts. She had also had nearly two years of experience with reform-inspired math in fourth and fifth grades. Her biggest challenge in class was participating in whole class groupings. She understood that she needed to participate but was not always comfortable doing so. For Anna, lack of math skill was not a fundamental problem that impeded her learning. She was becoming accustomed to the ways of school and the way in which math was taught at Ritter (CMP math), but there was a disconnect between her participation style and the expectations of Ms. Koch. In Ms. Koch's class, being quiet and reserved (during whole class and group work) was not recognized as a valuable quality. While Anna understood how she needed to act, she was not always successful in carrying it out.

Like Thomas and Pedro, Jennifer and Nestor were educated in the U.S. They were also familiar with reform math. What they lacked that the successful students had was math skill. More than once, I observed Nestor asking (anyone who would listen) the answers to simple multiplication and division problems. He used his fingers very often to help add and subtract simple problems, and answered his teacher incorrectly when she asked the class questions that required basic skill knowledge such as multiplication tables. Jennifer also struggled with basic mathematical problems which involved multiplication and division. Her strengths were in interacting in various group settings. However, she appeared to have a fundamental lack of basic skills.

Like Anna, Benjamin had only had a few years of schooling in the U.S. However, he was quite proficient in adhering to the rules of middle school. Although he was somewhat of a taciturn and reserved student, Benjamin made it clear that he wanted to be considered a 'regular student'. He came to class on time, was usually prepared, laughed at others' jokes at the right moments, answered teacher questions when called upon (but was not always right), turned in his work when requested to, and generally played the part of a regular American sixth grader. But like Jennifer and Nestor, Benjamin was lacking some critical mathematical skills to carry out the tasks assigned to him. It was not crystal clear, however, how much mathematical knowledge he was lacking, because his English was rather limited.

Later in the semester, both Benjamin and Nestor opened up. They began to participate more in varied groupings in class and were both much more successful at answering teacher questions correctly and finding answers to mathematical tasks assigned to them through the text or by the teacher. Similarly, Anna began to interact and participate in greater and more varied ways later in the semester. These three students clearly benefitted from extended time in the math classroom. They became more familiar with not only the ways of interacting in math class, but also with the concepts they were learning and reviewing throughout the semester. More time and more practice helped these learners. Unfortunately, it was a little too late for Nestor and Benjamin, who would have benefitted from extra attention and more time earlier in the semester.

Language Proficiency

Limited language proficiency in English was an undeniable barrier for learning math in CMP for the two recent arrivals, Anna and Benjamin. Both were highly driven and motivated to learn English, and both used English often in math. Although Anna was a beginning learner of English, she worked hard to overcome obstacles in school. She worked with friends, enlisted her sister's help, and studied a great deal to succeed. One of her primary goals was to return to Mexico. She told me that learning English was one of the main accomplishments she hoped to achieve before returning. Her test scores in reading and other content areas indicated that she was advancing rapidly.

Benjamin was so driven to learn English that he would not use his L1 in school. Standardized assessments indicated that he, like Anna, was also progressing quickly in English. Despite his teacher's ability to communicate with him in his native language, Benjamin did not have the support systems others like Anna had in his math class. Since Benjamin chose not to use his first language in math, he cut off that line of aid that could have served to elucidate important concepts.

Limited language proficiency was also a problem for Nestor. His test scores and grades did not show the impressive improvement that the immigrants Anna and Pedro showed. Although he had attended school in the U.S. throughout his life, he was still exempt from standardized tests as late as fifth grade. Like the immigrants, Nestor was receiving language support in the form of ESL, however, his language needs were clearly not being met. When I asked him what he learned in ESL he replied: "They show you stuff in Spanish and English . . . um, you read, Spanish and English, and they give you all these uh, DOL (daily oral language) so you can learn more about um, the letters, how do they pronounce them."

When I pressed him as to whether his ESL class was actually conducted often in Spanish, he said yes. English language learning was a problem for Nestor, however, a combination of factors were converging to endanger his success in school. He preferred to work alone (and was allowed to at times), or with English monolinguals (his favorite, being Scott, the special education student who was mainstreamed in Ms. Loesely's third period). In addition, Nestor was absent from school quite a lot. His frequent absence from class certainly contributed to his low performance in school. Nestor was a student who had begun to fall though the cracks.

Frames and Focus on Math Discourse: How CMP Supports Participation

Building background knowledge and creating connections is critical for students to be able to use their prior knowledge and apply it to the knowledge to be learned. Approaches designed for teaching ELLs such as Sheltered Instruction and *CALLA* (Chamot & O'Malley, 1994) advocate the building of background knowledge for learners within academic curriculum.

According to *The SIOP model*, a controversy exists between teaching students through their home or cultural background and 'adolescent' culture, or American pop culture (Echevarria et al., 2008). Between these two perspectives for teaching adolescent language learners, CMP sides with the latter. As shown in Chapter 4, mathematical concepts and problems in CMP are framed within interesting, relevant, and motivating situations that sixth graders from varying cultural and ethnic backgrounds, and from any number of American cities or towns might relate to. However, frames of play must not only be motivating to be effective, they must be taught in a manner that will promote connection building between an ELL or bilingual student's background knowledge.

Frames Support Participation

Frames of play used in CMP lessons (addressed in chapter four) serve to increase motivation and participation of students. The frames used in CMP invoke students' background knowledge about particular mathematical topics and strive to situate learning in a context that is appropriate and relevant for them. Attempting to target the interests of such a wide audience is a formidable task. While in many cases, the frames served their purpose of engaging learners and promoting learning, in others, students were presented with further complications as a result of their lack of background knowledge within the frame. A risk in utilizing frames of play to situate mathematical concepts is that students do not see the connections between the math and the frame. In other words, some students may not be as quick as Thomas (Example 4.1) when he said "I know how it goes!", but may respond as Jennifer did when she was asked in an interview what she had been learning in math, and replies "brownies." Her comment suggested that she did not see the connections between doubling and tripling recipes to learn about fractions.

While frames can be useful to facilitate student engagement, participation, and expanded learning roles through games, motivation and background knowledge, there is the potential for clouding learning for some. As Ms. Loesely explained, the students in her classes had never done activities such as making brownies from scratch, attended summer camp, or planned and planted a garden with their families. She argued that they lacked the background knowledge to properly comprehend the mathematical tasks situated within frames in CMP. But in examining focal students' discourse and written assignments when interacting within framed lessons, I found that they were able to negotiate through difficulties brought on by lack of background knowledge by talking with each other or by asking the teacher for assistance. When Jennifer and Kristina were increasing the brownie recipe to make enough for the entire sixth grade class (Example 3.8) they came across ingredients they weren't familiar with, such as condensed milk. Because they worked together and had their teacher as a resource, they were able to find answers to satisfy the problems they encountered. Jennifer was fortunate to have a support system of bilingual friends to help her out when she lacked the background knowledge, context, or language related problems.

Math Discourse

The CMP curriculum stresses the importance of mathematical discourse for critical thinking and learning. However, as shown in chapter five, a clear outline of what mathematical discourse entails is not always present in the curriculum. To encourage students to be successful in their use and development of mathematical discourse and accompanying mathematical concepts, the teachers had to delineate the discourse for their students and help them comprehend what the critical mathematical discourse is—in terms of vocabulary, syntax, semantics, and discourse patterns, as well as with respect to the complex higher order language skills and language functions relevant to mathematics. Some of the Springvale teachers focused on this, but others did not. Of the three teachers, Ms. Koch spent the most time working on math discourse. Overall, I found that when the teachers provided opportunities for participation in mathematical discussions, students had more opportunities to experiment with and practice math discourse, and to ultimately gain ownership of the complex, situationally-specific academic language valued in math class.

Other Features of the CMP Classes that Support Participation

Lack of background knowledge for some of the scenarios, or frames of play, put forth in the CMP text may play a role in the Latina/o ELLs' misunderstanding of math. However, Ms. Koch did not believe that the lack of background knowledge by those Latina/o, inner city, low SES sixth graders in her class was a fundamental problem. The grouping structure of Ms. Koch's class (which was built into CMP) served to support and facilitate learning and prevented potential roadblocks. For Anna, bilingual Pedro, as well as some of the other bilinguals in class helped fill in the gaps that her lack of knowledge left. In addition, throughout much of the spring semester in Ms. Koch's class, there was a total of three skilled adults (teacher, assistant, and student teacher) to assist learners in math. (In Ms. Loesely and Mr. Martinez' classes, there was only the teacher.) Ms. Koch effectively targeted student attention, motivation, and participation through adolescent culture. She advocated problem solving and negotiation in groups, and consistently provided lessons framed as play.

When she used CMP, Ms. Loesely was true to its structure and presentation. She utilized manipulatives, including blocks, calculators, colored rods and shapes, food, and money. She encouraged students to work together, and provided additional resources that would help students develop mathematical discourse and content.

In Mr. Martinez' class, his language learners had the benefit of having a teacher who not only spoke their native language, but who also intimately knew the community in which they lived. Benjamin and Thomas, as well as the other ELLs and bilinguals in their class, received the extra bonus of learning mathematics within a frame of reference they could easily understand. Mr. Martinez also provided expert scaffolding for some students in his class. Thomas was a common recipient of his teacher's expertise and attention. One significant difference between Mr. Martinez' class and the others, was that Mr. Martinez did not utilize manipulatives, group work, or as many lessons framed within motivating, interesting scenarios. Therefore, while Mr. Martinez' style of teaching closed some of the gaps in background knowledge that his learners had, other gaps still remained open.

Expert resources were uneven across the three classes, however, many of the focal students in each class had opportunities to work with experts. With sufficient expert resources to assist in scaffolding, the novice learners in all three of the sixth grade CMP classes were able to negotiate through potential problems and gaps in their background knowledge. Thus, potential difficulties in CMP can be avoided with expert resources and sufficient opportunities to interact with those resources. In sum, tailoring student learning to one's background knowledge can be beneficial, as it was in the three sixth-grade classes at Springvale ISD. However, additional measures must be taken to ensure that gaps in learning are closed. Scaffolding and native language support were effective measures for some focal students (who received it), in particular, Thomas, Pedro, Anna, and Jennifer.

CMP, Math Skills, and Assessment

While a common complaint by those familiar with CMP is that it does not teach skills crucial for success in math class, teachers themselves developed ways of handling the shortfalls inherent in CMP. Mr. Martinez, himself a critic of CMP, felt that the curriculum lacked important concepts that sixth graders needed to know such as how to add, subtract, divide, and multiply fractions. As a result, he chose not to use the CMP texts much. Instead, he opted for the more straightforward (non reform-based) Glencoe text (Boyd et al., 2001) and consumable worksheets, which were usually devoid of language and rife with numbers and symbols. Ms. Loesely agreed that there was not enough practice in CMP. She augmented her lessons on occasion with worksheets. When students failed exams or did not turn in assignments, she called them to her class at lunch, to make up the work. Where Ms. Koch also agreed (to a lesser extent than the Hot Springs teachers) that there were indeed some holes in the curriculum in terms of basic skills, she filled them by conducting whole class warm-up activities, assigning homework, taught class during advisory, and held weekly tutoring sessions. In general, though, Ms. Koch was a true proponent of the CMP curriculum.

Because CMP was aligned with the state standardized math test, Ms. Koch was preparing her students for tests daily. She also incorporated warm-ups and activities that followed the format of the state test. In addition, Ms. Koch gave pretests under the orders of the principal, and preparation worksheets as many teachers did, but her use of the CMP curriculum predominated. At Hot Springs, Mr. Martinez and Ms. Loesely used worksheets and state test practice booklets in class to prepare their students for state tests. As the date for the state standardized test neared, Mr. Martinez and Ms. Loesely's students did more and more state test worksheets, and less and less CMP. In double math, they prepared for the state test by working in their test preparation booklets and used district-specified problem solving strategies to solve math problems that resembled test questions. Unlike Ms. Koch, Mr. Martinez and Ms. Loesely did not feel that CMP was a good preparation for the state test—despite the fact that the district had formed a committee of math teachers to clearly delineate the connections between CMP and the state tests as well as with the state's essential elements, or critical knowledge and skills. In the end, the district math state test pretests showed that Ms. Koch's classes had some of the highest scores in the district, while Mr. Martinez had some of the lowest. CMP was a beneficial test preparation, as well as a useful curriculum for teaching math skills for the sixth graders at Springvale Independent School District.

INTERNAL AND EXTERNAL FACTORS AFFECT PARTICIPATION

Both internal factors related to the individual, and external factors related to the CMP curriculum affect ELLs' participation in reform math. When factors related to the individual, such as language proficiency, gender, ethnicity, and knowledge of math and school are considered within the curriculum, language learners' development is facilitated. As mentioned, aspects of the presentation of CMP are similar to the goals promoted in ESL pedagogy such as Sheltered Instruction. Thus, reform mathematics, and CMP in particular, can be a very effective curricular approach for learners of English.

As seen in chapters three, four, and five, an appropriate learning environment that fosters cooperation through both the L1 and the L2 of the students can facilitate learning. This environment includes an explicit focus on mathematical discourse, an understanding of language learners' special needs with respect to cultural and familial issues (Benjamin's rejection of Spanish in school, Anna's shy demeanor, Nestor's absenteeism and transience), as well as an emphasis on creating and building critical background knowledge, and providing scaffolding when necessary to clarify concepts. The focus on group work, learning strategies, and higher order thinking within a learning environment that utilizes hands-on activities framed within motivating scenarios is beneficial for language learners. Like sheltered instruction, CMP can help make mathematical content as well as mathematical language comprehensible. When used appropriately, CMP is an effective curriculum for ELLs.

THE EFFECT OF PARTICIPATION ON LEARNING AND SOCIALIZATION

Overall, I found that some of the focal students in my study developed proficiency in mathematical discourse and enjoyed success in traditional measures of academic progress such as standardized tests and grades. Thomas, Pedro, and Anna were ultimately successful in school, despite the fact that Anna was often inconsistent in her participation. On the other hand, Jennifer often participated in varied group structures, displayed both higher order thinking skills, as well as sophisticated learning strategies, but was ultimately not successful in terms of school grades and test scores. Those students who were not successful in using the math discourse required for participation in fraction lessons, namely, Jennifer, Nestor and Benjamin, did not achieve better than average grades or pass classroom or standardized tests, although some participated fully as peripheral or integral participants in classroom activities.

The less successful students, Nestor, Benjamin, and Jennifer, would certainly have benefitted from richer opportunities for academic participation, but participation alone cannot explain their lack of success. Certainly, a language learner may have linguistic difficulties in the target language, English, but they can still participate actively and engage within the community of practice in their classroom. Jennifer was a prime example of this. She engaged with others, shared in the enterprise and repertoire of knowledge, (and in some cases, discourse) utilized higher order thinking and learning strategies as a good learner would, but still failed in some of the critical areas of mathematics. In some cases, students ultimately labeled as 'unsuccessful' played diverse roles in participation frameworks created in their classes, shifting between expert and novice as they experimented with knowledge. The issue of learner agency complexifies and problematizes the notion that learners, and in particular, (novice) language learners can participate (fully) and socialize each other to new knowledge, when they themselves have not attained the designation of expert (see Schecter & Bayley, 2004).

While an understanding of participation, engagement, and identity helps show how discourse practices evolve in math class, it is vitally important that other factors are taken into consideration. Language learning through content in school is a multifaceted, complex process and as a result, gaining an understanding of the process entails consideration of many issues.

IMPLICATIONS

The implications I draw from this study are varied. They deal in particular with mathematical discourse, the CMP curriculum, and education and training of mathematics teachers of ELLs, with a specific emphasis on the teacher's classroom structure and methods of interaction with students. As reported in chapter five, mathematical discourse is not spelled out clearly for teachers, and much less for students. Teachers may be unaware of the situated syntax, semantics, and lexicon used in math lessons. As a result, teachers need to highlight and emphasize the important aspects of mathematical discourse that students are responsible for. An explicit focus on discourse in the classroom will help students to become socialized to the language and concepts they need to know. Students will then be more likely to take ownership of the discourse and use it to socialize each other.

An explicit focus on discourse in the classroom can occur in many ways. Like Mr. Martinez, teachers can use the students' L1 to highlight important concepts and terms. Also, teachers can use bilingual grouping, as Ms. Koch did, to help students help themselves learn mathematical discourse. In addition, teachers should incorporate literacy into the learning of math discourse. Preparing wall charts, writing on the board, encouraging students to create and use dictionaries, or teaching explicit knowledge of cognates are other strategies that can be utilized to promote focus on mathematical discourse.

As reported by the three teachers, CMP leaves gaps (related to skill) that other sixth grade mathematics curricula may not. Thus teachers themselves need to have the freedom (and expertise) to fill in the gaps for their students, but at the same time they must provide rich opportunities for interaction, discussion, scaffolding, and socialization. All three of the teachers in my study did in fact do this, but Ms. Koch was the only one who did so by keeping CMP as the primary curriculum in her class. She assigned warm-up activities, homework, and outside assignments to her students to practice skills and take grades. However, she reserved class time for students to practice concepts using interactive language, manipulatives, and motivating tasks framed in interesting ways. When Ms. Loesely and Mr. Martinez felt the need for 'skill and drill', they turned to traditional textbooks and worksheets. Since they weren't 'allowed' to have warm-ups in class, they turned to other means that often align with the requirements of the state standardized tests. Ms. Loesely presented traditional activities in innovative ways, incorporating her own personal hands-on materials into the activities, or grouped students to work cooperatively. In contrast, students in Mr. Martinez' class were expected to work as a whole class or individually. As a result, some opportunities for rich, discursive interaction were lost. To be sure, teachers are responsible for determining where what and when curriculum should be implemented in their own classrooms, however, it is important for them to do so by providing the richest opportunities for student interaction.

When using reform-oriented curricula such as CMP with language learners, teachers must not only be properly trained in CMP, but should also be aware of specific needs of ELLs and how the curriculum can help serve those needs. Teachers should be knowledgeable of potential language resources that can be used to facilitate discourse learning. Although all three of the teachers in my study had additional resources in Spanish such as mathematical dictionaries, none was aware that there were Spanish language resources that accompanied the CMP texts to facilitate transition of mathematical learning from Spanish into English. Resources such as this would be exceptionally useful for newcomers, or as a supplementary resource to the primary English materials for intermediate or advanced ELLs.

Teachers must understand that language learners may experience affective issues that cause them to reject or ignore their specialized language needs. Benjamin, for example, resisted use of Spanish in school. Although Mr. Martinez continued to address him in Spanish, Benjamin persisted in using English (and rejected Spanish)—even when it may have helped him become more successful. Teachers need to be aware of such issues and work to discover reasons why their students do not succeed.

Similarly, teachers need to consider all learning styles, and recognize that students from other cultures will come to the classroom with different notions of classroom interaction. In Ms. Koch's class, Anna was often quiet and reserved. While she was praised for her bilingualism, Anna's teacher often reproached her for staying quiet. A non-traditional mathematics class in which candy is eaten, spinners are spun, and students are working in groups daily has the potential to challenge even an American student's traditional idea of math class. For a learner coming from a culture in which math is a quiet, individual, teacher-centered discipline (such as immigrants Anna and Benjamin), there is an even greater possibility for confusion. When there is such a discontinuity between teacher beliefs and student learning style (or student notions of how mathematics class *should* be conducted) potential difficulties can occur.

In an interactive, group oriented academic class, appropriate grouping matters. For ELLs and bilinguals who lack language, background, cultural knowledge, or mathematical skill, participating within a heterogeneous group who can scaffold learning is crucial. Anna's interactions with Pedro helped her learn math through language. Similarly, the problem of Jennifer's lack of background knowledge was alleviated with assistance from her friends and teacher. Incorporating and building on students' background knowledge, as it is in CMP through framing of lessons, is useful—but it is important to help learners see the connection to math. Well-structured grouping strategies can help students see the connections and receive the linguistic and other assistance they need in school.

Finally, while I did not find that gender was a critical issue in students learning at sixth grade, teachers need to acknowledge that it could be an issue at later grades. Middle school teachers need to keep in mind that success in math by adolescent Latinas is confounded by a number of factors. In particular, culture, home responsibilities, and language, are just a few of the factors that can impede or promote success in mathematics.

LIMITATIONS OF THIS STUDY AND FURTHER RESEARCH

In order to make conclusions based on a wider population of adolescent Latina/o language learners, a wider sample of students should be analyzed over a longer period of time. While ethnographic case studies can show learning of individuals and can suggest implications based on their analyses, they cannot be used to generalize over a population.

My study followed in detail six language learners in the second half of their sixth grade year. Limitations of my study include not only the (relatively short) time in which I followed them, but also the lack of some data sources. There were certain gaps in my data. When collecting student records and student assignments to triangulate my analyses, I relied on overworked and underpaid school administrative assistants and sixth grade students to provide data for me. In addition, I was unable to remunerate my participants. I was thus dependent on their kindness and flexibility to assist me in my data collection.

My decision to focus on only two students per class was one I made later, when analyzing data. Had I made that decision when carrying out my fieldwork, I would have been able to collect, and ultimately analyze a much more concentrated sample of student discourse produced over the semester. However, I originally chose more students as I was concerned about collecting a representative sample of language learners, as well as with the problem of participant attrition.

In developing further research examining ELLs in reform-based mathematics classes I would follow a larger sample of students throughout middle school. I would include pre and post tests of math skill level and mathematical discourse and would be much more diligent in collecting student records and assignments. The present study examined a host of general questions regarding learning in reform-mathematics classes. Further research would hone in on specific issues I found to be relevant to language learners. I would examine in more depth the effect of background knowledge as related to the curriculum, the interplay between teaching style (and curriculum presentation) and learning style, as well as the students' reaction to and relationship to reform mathematics. Finally, I believe that a longitudinal comparison of language learning students in both reform-inspired and traditional mathematics classes would reveal important insights. A study such as this may also encompass ELLs and bilinguals from diverse, heterogeneous linguistic backgrounds—rather than just one. Such as study would show how the interplay of linguistic, educational, cultural, and mathematical issues affects language learners in school.

REFERENCES

American Community Survey. (2005). U.S. Bureau of the Census. Retrieved October 2008 from: http://www.census.gov/acs/www/index.html

Anderson, L., & Krathwohl, D. (Eds.). (2001). *A taxonomy for learning, teaching and assessing: A revision of Bloom's Taxonomy of educational objectives* (Complete ed.). New York: Longman.

Batalova, J., Fix, M., & Murray, J. (2005). *English language learner adolescents: Demographics and literacy achievements.* Report to the Center for Applied Linguistics. Washington, DC: Migration Policy Institute.

Bateson, G. (1972). A theory of play and fantasy. Reprinted in *Steps to an ecology of a mind* (pp. 117–193). New York: Ballantine Books.

Bayley, R. (2004). Linguistic diversity and English language acquisition. In E. Finegan, & J. Rickford. (Eds.), *Language in the USA: Themes for the 21st century* (pp. 268–285). Cambridge, UK: Cambridge University Press.

Bloom, B., Englehart, M., Furst, E., Hill, W., & Krathworl, D. (Eds.). (1956). *Taxonomy of educational objectives: The classification of educational goals.* Handbook I: *Cognitive domain.* New York: David McKay Co.

Boyd, C., Moore-Harris, B., Howard, A., & Molina, D. (2001). *Mathematics: Applications and connections: Course 1.* Westerville, OH: Glencoe-McGraw Hill.

Brilliant-Mills, H. (1994). Becoming a mathematician: Building a situated definition of mathematics. *Linguistics and Education 5*, 301–334.

Capps, R., Fix, M., Murray, J., Ost, J., Passel, J., & Herwantoro, S. (2005). *The new demography of America's schools: Immigration and the No Child Left Behind Act.* Washington, DC: The Urban Institute. Retrieved October 2007, from: http://www.urban.org/publications/311230.html

Chamot, A., & O'Malley, M. (1994). *The CALLA handbook: Implementing the cognitive academic language learning approach.* Reading, MA: Addison Wesley Publishing.

Corsaro, E. (1988). Routines in the peer culture of American and Italian nursery school children. *Sociology of Education*, 1–14.

English Language Learners and Math, pages 137–141

Cummins, J. (1994). Knowledge, power, and identity in teaching English as a second language. In F. Genesee (Ed.). *Educating second language children* (pp. 33–58). New York: Cambridge University Press.

Diaz-Rico, L. (2008). *A course for teaching English language learners.* Boston: Allyn & Bacon.

Echevarria, J., Vogt, M., & Short, D. (2004/2008). *Making content comprehensible for English learners: The SIOP model.* Boston: Pearson Education.

Echevarria, J., & Graves, A. (2007). *Sheltered content instruction: Teaching English language learners with diverse abilities.* Boston: Pearson Education.

Garrett, P., & Baquedano-López, P. (2002). Language socialization: Reproduction and continuity, transformation and change. *Annual Review of Anthropology 31,* 339–361.

Garvey, C. (1977). *Children's play.* Cambridge, MA: Harvard University Press.

Gee, J. (2005). *An introduction to discourse analysis: Theory and method* (2nd ed.). London: Routledge, Taylor and Francis.

Geertz, C. (1973). *The interpretation of cultures.* New York: Basic Books.

Gibbons, P. (1993). *Learning to learn in a second language.* Portsmouth, NH: Heinemann.

Gilligan, C. (1982). *In a different voice.* Cambridge, MA: Harvard university Press.

Goffman, E. (1974). *Frame Analysis.* New York: Harper and Row.

Goffman, E. (1983). Frame analysis of talk. From 'Felicity's condition'. *American Journal of Sociology 89*(1), 1–3, 25–51.[Reprinted in *The Goffman Reader.* (1997). Lemert, C. & Branaman, A. (Eds.). Malden, MA: Blackwell Publishers.]

Gumperz, J. (1982). *Discourse strategies.* Cambridge: Cambridge University Press.

Halliday, M. (1978). *Language as a social semiotic.* Baltimore, MD: University Park Press.

Hansen-Thomas, H. (2008). The math initiative in a 7th grade science class: How a daily routine results in academic participation by ELLs. In B. K. Richardson & K. Gomez (Eds.), *The work of language in multicultural classrooms: Talking science, writing science* (pp. 377–413). Mahwah, NJ: Lawrence Erlbaum.

He, A. (2003). Novices and their speech roles in Chinese heritage language classes. In R. Bayley & S. R. Schecter (Eds.), *Language socialization in bilingual and multilingual societies* (pp. 128–146). Clevedon, UK: Multilingual Matters.

Hoyle, S. (1993). Participation frameworks in sportscasting play: Imaginary and literal footings. In D. Tannen (Ed.), *Framing in discourse* (pp. 114–145). Oxford: Oxford University Press.

Hymes, D. (1974). Ways of speaking. In R. Bauman & J. Sherzer (Eds.), *Explorations in the ethnography of speaking* (pp. 433–451). Cambridge: Cambridge University Press.

Institute of Education Sciences. (2007). *Status and trends in the education of racial and ethnic minorities.* Retrieved October 2008 from: http://nces.ed.gov/pubs2007/minoritytrends/ind_1_2.asp#f5

Kanagy, R. (1999). Interactional routines as a mechanism for L2 acquisition and socialization in an immersion context. *Journal of Pragmatics, 31,* 1467–1492.

Kanno, K. (1999). Comments on Kelleen Toohey's "Breaking them up, taking them away: ESL students in grade 1" The use of the community of practice perspective in language minority research. *TESOL Quarterly, 33*(1), 126–132.

Krashen, S. (1982). *Principles and practices in second language acquisition.* Oxford: Pergamon Press.

Lakoff, G. (2004). *Don't think of an elephant: Know your frames and frame the debate: The essential guide for progressives.* White River Junction, VT: Chelsea Green Publishing Co.

Lantolf, J. (2000). Introducing sociocultural theory. In J. Lantolf (Ed.), *Sociocultural theory and second language learning* (pp. 1–26). Cambridge: Cambridge University Press.

Lappan, G., Fey, J. T., Fitzgerald, W. M., Friel, S. N., & Phillips, E. D. (1996). *A guide to the connected mathematics curriculum: Getting to know connected mathematics.* White Plains, NY: Dale Seymore Publications.

Lappan, G., Fey, J. T., Fitzgerald, W. M., Friel, S. N., & Phillips, E. D. (1998). *Connected* Mathematics, Michigan State University: How likely is it? Probability. White Plains, NY: Dale Seymore Publications.

Lappan, G., Fey, J. T., Fitzgerald, W. M., Friel, S. N., & Phillips, E. D. (2002a). *Connected Mathematics, Michigan State University: Bits and Pieces I, Understanding rational numbers.* Glenview, IL: Prentice Hall.

Lappan, G., Fey, J. T., Fitzgerald, W. M., Friel, S. N., & Phillips, E. D. (2002b). *Connected Mathematics, Michigan State University: Bits and Pieces II, Using rational numbers.* Glenview, IL: Prentice Hall.

Lave, J. (1991). Situating learning in communities of practice. In L. Resnick, J. Levine, & S. Teasley (Eds.), *Perspectives on socially shared cognition* (pp. 63–84). Washington, DC: American Psychological Association.

Lave, J., & Wenger, É. (1991). *Situated learning: Legitimate peripheral participation.* Cambridge: Cambridge University Press.

Lytra, V. (2004). Frame shifting and identity construction during whole class instruction: Teachers as initiators and and respondents in play frames. In M. Baynham, A. Deignan, & G. White (Eds.), *Applied linguistics at the Interface* (British Studies in Applied Linguistics, Vol. 19) (pp. 120–131). London: Equinox.

Lytra, V. (2008). Playful talk, learner's play frames and the construction of identities. In M. Martin-Jones & A-M. De Mejia (Eds.), *Encyclopedia of language and education,* Vol. 3: *Discourse and education.* New York: Springer.

MacWhinney, B. (2000). *The childes project: Tools for analyzing talk. Volume 1: Transcription format and programs.* Mahwah, NJ: Erlbaum.

Merriam, S. (1998). *Qualitative research and case study applications in education.* San Francisco: Jossey-Bass.

Molesky, J. (1988). Understanding the American linguistic mosaic: A historical overview of language maintenance and language shift. In S. McKay & S. Wong (Eds.), *Language diversity: Problem or resource.* Boston: Heinle & Heinle.

National Clearinghouse for English Language Acquisition and Language Instruction Education Programs (2006). Retrieved October 2008 from: http://www.ncela.gwu.edu/stats/

National Council of Teachers of Mathematics. (2000). *Principles and standards for school mathematics.* Reston, VA: NCTM.

National Council of Teachers of Mathematics. (1991). *Professional standards for teaching mathematics.* Reston, VA: NCTM.

O' Connor, M., & Michaels, S. (1996). Shifting participant frameworks: Orchestrating thinking practices in group discussion. In D. Hicks (Ed.), *Discourse, learning, and schooling* (pp. 63–103). Cambridge: Cambridge University Press.

Peregoy, S., & Boyle, O. (2008). *Reading, writing, and learning in ESL:A resource book for teaching K–12 English learners* (5th ed.). Boston: Pearson Education.

Peterson, D. (Oct. 2004). Definition of viniculum. The math forum at Drexel University: Ask Dr. Math. Retrieved June 2, 2009 from http://mathforum.org/library/drmath/view/69150.html.

Pinnell, G. S. (1985). Ways to look at the functions of children's language. In A. Jaggar & M. T. Smith-Burke (Ed.), *Observing the language learner* (pp. 57–72). Newark, DE: International Reading Association.

Rampton, B., Roberts, C., Leung, C. & Harris, R. (2002). Methodology in the analysis of classroom discourse. *Applied Linguistics, 23*(3), 373–392.

Rogoff, B. (1990). *Apprenticeship in thinking: Cognitive development in social context.* New York: Oxford University Press.

Rumelhart, D. (1975). Notes on a schema for stories. In D. Bobrow & A. Collins (Eds.), *Representation and understanding* (pp. 211–236). New York: Academic Press.

Sadker, M., & Sadker, D. (1995). *Failing at fairness: How America's school's cheat girls.* New York: Touchstone Press.

Schecter, S., & Bayley, R. (2004). Language socialization in theory and practice. *International Journal of Qualitative Studies in Education, 17*(5), 605–625.

Schieffelin, B., & Ochs, E. (1986). Language socialization. *Annual Review of Anthropology, 15,* 163–191.

Schiffrin, D. (1994). *Approaches to discourse.* Cambridge, MA: Blackwell.

Schiffrin, D., Tannen, D., & Hamilton, H. (Eds.). (2001). *The handbook of discourse analysis.* Malden, MA: Blackwell.

Sfard, A. (1998). On two metaphors for learning and the dangers of choosing just one. *Educational Researcher, 27*(2), 4–13.

Tannen, D. (1993). What's in a frame? Surface evidence for underlying expectations. In D. Tannen (Ed.), *Framing in discourse* (pp. 14–56). Oxford: Oxford University Press.

Tannen, D., & Wallat, C. (1993). Interactive frames and knowledge schemas in interaction: Examples from a medical examination/interview. In D. Tannen (Ed.), *Framing in discourse* (pp. 57–74). Oxford: Oxford University Press.

U.S. Department of the Census. (2000). Author. Retrieved October 2008 from: http://www.census.gov/main/www/cen2000.html

U.S. Bureau of the Census. (2003). *File QT-P16: Language spoken at home: 2000,* Washington, DC: Department of Commerce.

Varghese, M., Morgan, B., Johnston, B., & Johnson, K. (2005). Theorizing language teacher identity: Three perspectives and beyond. *Journal of Language, Identity, and Education, 4*(1), 21–44.

Vygotsky, L. S. (1978). *Mind in Society.* Cambridge, MA: Harvard University Press.

Wenger, E. (1998). *Communities of practice: Learning, meaning, and identity.* Cambridge: Cambridge University Press.

Wertsch, J. (1991). A sociocultural approach to socially shared cognition. In L. Resnick, J. Levine, & S. Teasley (Eds.), *Perspectives on socially shared cognition* (pp. 85–100). Washington, DC: American Psychological Association.

Wohlwend, K. (2007). More than a child's works: Framing teacher discourse about play. *Interactions: UCLA Journal of Education and Information Studies, 3*(1,4), 1–27.

Willett, J. (1995). Becoming first graders in an L2: An ethnographic study of L2 socialization. *TESOL Quarterly, 29,* 473–503.

ABOUT THE AUTHOR

Holly Hansen-Thomas is an Assistant Professor of English as a Second Language and Bilingual Education at Texas Woman's University in Denton, Texas. Dr. Hansen-Thomas is proud to be a recently-repatriated native Texan serving the students and teachers of her home state.

Dr. Hansen-Thomas has been involved in second language education for nearly two decades. She has taught both English as a Second Language and English as a Foreign Language in a variety of settings, including public schools, private language institutions, colleges, and universities throughout the U.S. and the international community. She has taught and studied in Budapest, Hungary, as a Visiting Professor on a Fulbright Fellowship, as well as in Barcelona, Spain; Dresden, Germany; Binghamton, New York and various locations in Texas.

Hansen-Thomas' scholarly work deals with issues of second language learners' teaching and learning in school. She is particularly interested in learning how adolescent ELLs develop academic language and literacy in mainstream classes such as mathematics and science. Dr. Hansen-Thomas has an active presentation record, having spoken on issues of language education at local, national, and international conferences throughout the U.S., as well as in Mexico, Hungary, and Belarus. Her published work includes articles in peer-reviewed academic journals, chapters in educational texts and reference books, reviews in print and online sources, and educational periodicals, among others. This is her first single-authored book.

English Language Learners and Math, page 143

CPSIA information can be obtained at www.ICGtesting.com
Printed in the USA
LVOW101016070213

319080LV00005B/31/P